# Biophysical Chemistry
## 2nd Edition

**Alan Cooper**
*Department of Chemistry, University of Glasgow, UK*

Tutorial Chemistry Texts No. 24

ISBN: 978-1-84973-081-5
ISSN: 2045-6611

A catalogue record for this book is available from the British Library

Published by The Royal Society of Chemistry,
Thomas Graham House, Science Park, Milton Road,
Cambridge CB4 0WF, UK

Registered Charity Number 207890

For further information see our web site at www.rsc.org

# Preface

Biology is chemistry on an impressive scale. It is a product of evolution—the outcome of countless random experiments, resulting in the exquisite complexity of the biological world of which we are a part. Setting aside any philosophical considerations, living organisms—including ourselves—are simply nothing more than wet, floppy bags of chemistry: complicated mixtures of molecules interacting in a multitude of ways. And all this takes place mainly in water, a solvent that most chemists try to avoid because of its complexities. But we can learn from this. In the course of evolution, biology has had the opportunity to perform vastly more experiments than we can ever contemplate in the laboratory. The resulting chemistry is fascinating in its own right, and we can quite rightly study it for its intellectual satisfaction alone. We can also, if we choose, apply what we learn to other areas of chemistry and to its applications in biomedical and environmental areas.

This book is about the physical chemistry of biological macromolecules and how we can study it. The approach here is unashamedly experimental: this is the way science actually works, and in any case we do not yet have the rigorous theoretical understanding perhaps found in more mature areas of chemistry. This is what makes it a fun topic, and why it poses fascinating challenges for both theoretical and experimental scientists.

The level adopted in this tutorial text should be suitable for early undergraduate years in chemical or physical sciences. However, since this interdisciplinary topic is often postponed to later years, the book will also act as a basis for more advanced study. Students in other areas of biological sciences might also appreciate the less intimidating approach to physical chemistry that I have attempted here.

For this second edition, in response to (mostly) generous comments from students and other colleagues, I have corrected a few howlers, updated some of the material and added a chapter on imaging methods, including a brief overview of protein crystallography and diffraction techniques.

My family and other animals have matured somewhat since the first edition and have gracefully tolerated my periods of grumpiness during the revision of this text. Thank you. I hope I am better now. I am also grateful to my students and other colleagues who have checked and corrected some of the material and particularly to Adrian Lapthorn, Nicola Meenan, Brian O. Smith and Steven Vance for expert assistance in areas where my ignorance was hard to disguise. I did not always follow their suggestions—so errors and infelicities are mine.

Alan Cooper
Glasgow

# Contents

# 1
# Biological Molecules

You don't need to know any biology in order to study biological molecules, but it does help to have some background.

### Aims

This chapter will briefly review the bare bones of biological (macro)molecules. By the end, and together with your previous knowledge and some background reading, you should be able to:

- Describe the basic chemical structures of polypeptides, polynucleotides, fats, lipids, and carbohydrates
- Explain what is meant by the primary, secondary, tertiary and quaternary structure of proteins
- Describe the behaviour of fats, lipids and detergents in water
- Explain the anomalous properties of liquid water
- Recall the fundamentals of acid–base equilibrium

## 1.1   Introduction

This book is mainly about the experimental methods used to understand the physical properties and function of the molecules that make up living systems.

These molecules—proteins, polynucleotides, polysaccharides, lipids—are not necessarily any different from molecules we study in other branches of chemistry. But there are some additional factors, arising from their biological origin, which we need to be aware of.

- Biological macromolecules are large molecules formed from many smaller units and are (usually) polymers of precise length and specific sequence.
- They (usually) fold or associate into specific conformational assemblies stabilized by non-covalent interactions.
- This (usually) happens in water.
- The molecules are the (usually) successful outcomes of biological evolution.

It is this last point that makes things so exciting for the biophysical chemist. The molecules we see today are the results of countless random (more or less) experiments over millions of years during which living systems have evolved to take advantage of subtle principles of physical chemistry that we barely yet understand. By studying such systems we can learn much about physical chemistry in general, with potential for applications in other areas.

## 1.2    Proteins and Polypeptides

D-amino acids are encountered only in special instances such as bacterial cell walls and peptide antibiotics.

Proteins are polymers made up of specific sequences of L-amino acids linked together by covalent peptide (amide) bonds (Figure 1.1). Amino acids are chosen from a basic set of 20 building blocks differing in side chain (Figure 1.2), with occasional special purpose side chains made to order (*e.g.* hydroxyproline).

The term 'molecular weight' is not strictly accurate (why?) but is commonly used, especially in the older (biochemical) literature. The more correct terms are 'relative molecular mass (RMM)' (no units) or 'molar mass' (kg mol$^{-1}$ or g mol$^{-1}$). One **dalton** (1 Da) is equal to 1 amu (atomic mass unit).

Typical proteins range in polypeptide chain length from around 50 to 5000 amino acids. The average relative molecular mass of an amino acid is around 110, so proteins can have RMMs from 500 to 500 000 (0.5–500 kDa) or more—especially in multi-subunit proteins consisting of specific aggregates (see Table 1.1).

### Worked Problem 1.1

**Q**: How many molecules are there in a 1 mg sample of a protein of 25 000 RMM?

**A**: 25 000 RMM $\equiv$ 25 000 g mol$^{-1}$

$$1\,mg \equiv 1 \times 10^{-3}/25\,000 = 4 \times 10^{-8}\,mol$$
$$\equiv 4 \times 10^{-8} \times 6 \times 10^{23} = 2.4 \times 10^{16}\,molecules.$$

### Worked Problem 1.2

**Q**: In a 1 mg cm$^{-3}$ solution of proteins with RMM 25 000, roughly how far apart are the molecules on average?

**A**: Volume per molecule = 1 (cm$^3$)/2.4 $\times$ 10$^{16}$ = 4.2 $\times$ 10$^{-17}$ cm$^3$.

So each molecule might occupy a cube of side 3.5 $\times$ 10$^{-6}$ cm (cube root of the volume), or 35 nm.

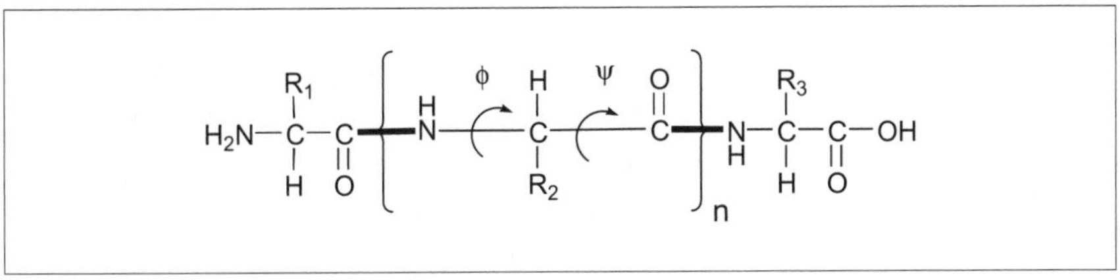

Figure 1.1 Polypeptide structure showing rotatable $\phi/\psi$ angles. The planar peptide (amide) bonds are shown in bold and are usually *trans*.

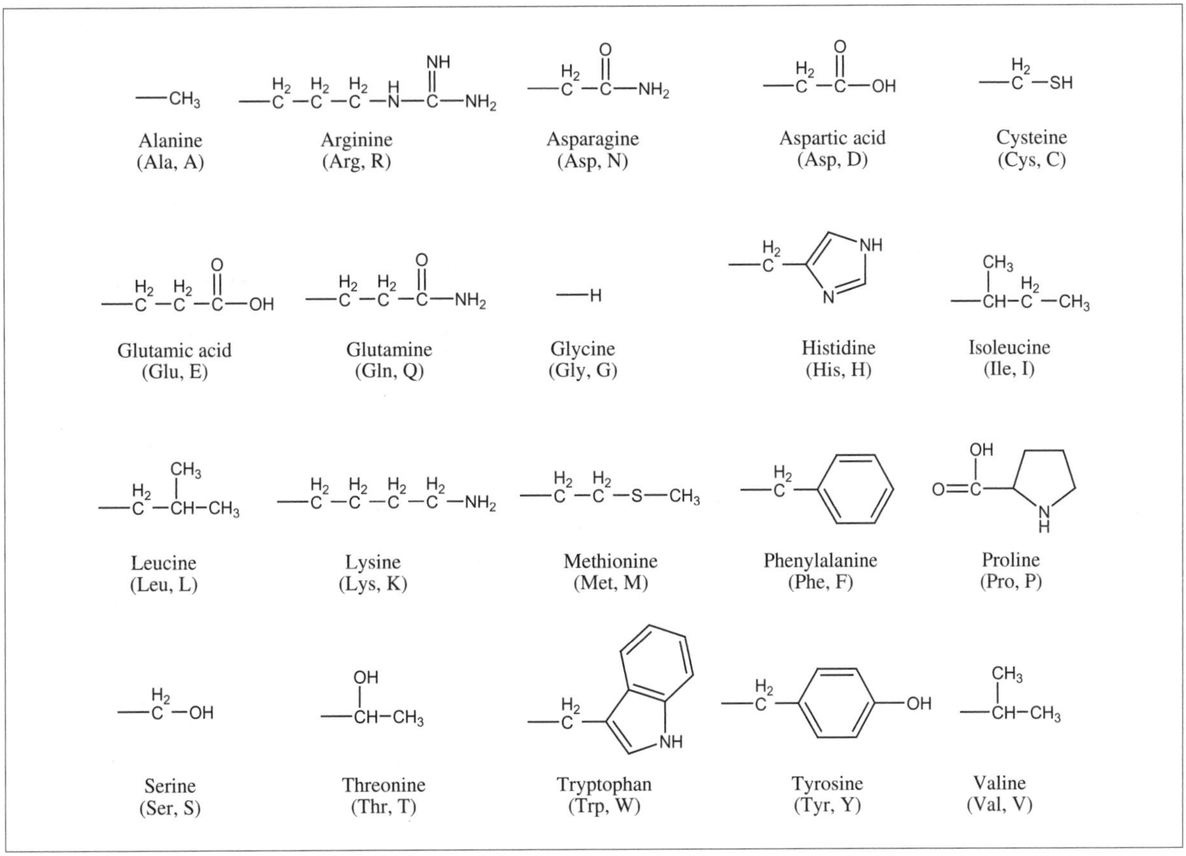

Figure 1.2 The 20 naturally occurring amino acid side chains (residues) with their three-letter and single-letter abbreviations.

---

## Worked Problem 1.3

**Q**: How does the answer to Worked Problem 1.2 compare to the size of one 25 000 RMM molecule?

**A**: Mass of 1 molecule $= 25\,000/6 \times 10^{23} = 4.2 \times 10^{-20}$ g.

This corresponds to a molecular volume of around $4.2 \times 10^{-20}\,\text{cm}^3$, assuming a density similar to water. This corresponds to a cube of side approximately 3.5 nm.

So in a $1\,\text{mg cm}^{-3}$ solution, these molecules are separated on average by about 10 molecular diameters.

**Table 1.1**   Some common proteins

| Name | No. of amino acids | RMM | Function |
|---|---|---|---|
| Insulin | 51 (2 chains, 21 + 30) | 5784 | Hormone controlling blood sugar levels. A-chain and B-chain covalently linked by disulfide bonds. Globular. |
| Lysozyme (hen egg white) | 129 | 14 313 | An enzyme that catalyses hydrolysis of bacterial cell wall polysaccharides. Found in egg white, tears and other biological secretions. Globular. |
| Myoglobin | 153 | 17 053 | Oxygen transporter in muscle. Contains haem group. Globular. |
| Haemoglobin | 574 ($2 \times 141$ + $2 \times 146$) | 61 986 ($2 \times 15\,126$ + $2 \times 15\,867$) | Oxygen transporter in blood stream. Consists of four subunits ($2\,\alpha$ and $2\,\beta$ chains), with haem. Globular. |
| Rhodopsin | 348 | 38 892 | Photoreceptor membrane protein in the retina of the eye. Contains 11-*cis* retinal as chromophore. |
| Collagen | 3200 (approx. $3 \times 1060$) | 345 000 | Connective tissue protein of skin, bone, tendon, *etc*. Three-stranded triple helix. Most abundant protein in animals. Fibrillar. |
| RuBISCO (ribulose bisphosphate carboxylase/oxygenase) | 4784 ($8 \times 475$ + $8 \times 123$) | 538 104 ($8 \times 52\,656$ + $8 \times 14\,607$) | Carbon fixation enzyme of green plants and algae; 16 subunits (eight large, eight small). Most abundant protein on Earth. |

Proteins function as enzymes (biological catalysts), antibodies, messengers, carriers, receptors, structural units, *etc.* Their chemical structure and molecular conformation are commonly described in terms of:

**Primary structure:** the sequence of amino acids in the polypeptide chain (see Figure 1.3). This is unique to each protein, and is determined (primarily) by the genetic information encoded in the DNA of the relevant gene.

**Secondary structure:** regular, repeating structures such as α-helix, β-sheets, *etc.* (see Figure 1.4).

**Tertiary structure:** the three-dimensional arrangement of secondary structure elements that defines the overall conformation of the (globular) protein (see Figure 1.5).

**Quaternary structure:** in multi-subunit proteins, the three-dimensional arrangement of the subunits (see Figure 1.6).

Because of rotational flexibility in the polypeptide backbone, primarily around the N–C$_\alpha$ ($\phi$) and C$_\alpha$–C ($\psi$) angles, there is a very large number of possible conformations that any one polypeptide molecule might adopt. Unlike most synthetic polymers, however, proteins have the ability to fold up (under the right conditions) into specific conformations and it is these conformations (structures) that give rise to their individual properties.

Haemoglobin, for example, is made up of four globular subunits—two of one kind (α) and two of another (β)—which combine to form a tetramer quaternary structure. Interaction between these subunits is responsible for the delicate control of oxygen uptake and release by the haem groups in this protein.

## Box 1.1 The 'Protein Folding Problem'

Most proteins do not have a problem folding—they just do it. However, we have a problem understanding how they do it and predicting what the conformation of a particular amino acid sequence will be.

```
KVFERCELAR TLKRLGMDGY RGISLANWMC LAKWESGYNT RATNYNAGDR
STDYGIFQIN SRYWCNDGKT PGAVNACHCS ASALLQDNIA DAVACAKRVV
RDPQGIRAWV AWRNRCQNRD VRQYVQGCGV
```

**Figure 1.3** Primary structure of a 130-residue protein (human lysozyme) shown using single-letter amino acid codes.

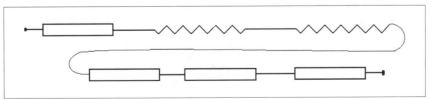

**Figure 1.4** Secondary structure.

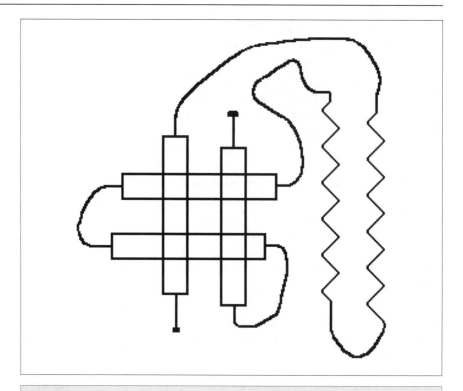

**Figure 1.5**   Tertiary structure.

You may be surprised to find that your calculator has trouble doing calculations such as $9^{100}$. Why? How can you get around it?

The complexity of the problem was highlighted some years ago by Cyrus Levinthal,[1] a computer scientist who was one of the first to tackle the problem.

Each $\phi$ or $\psi$ angle in a peptide might have roughly three possible values, giving $3 \times 3 = 9$ possible conformers for each peptide (not counting side chain conformers). For even a small polypeptide of 100 amino acids, this corresponds to at least $9^{100} \approx 3 \times 10^{95}$ possible different conformations of the polypeptide chain—only one of which (or a relatively small set) will be the 'correct' one.

Assuming (optimistically) that peptide conformations can switch on the femtosecond timescale ($10^{-15}$ sec), it would take a time of order $3 \times 10^{80}$ seconds or about $10^{73}$ years to search through all these possibilities to find the right one. This is a time much longer than the known age of the Universe. Yet proteins actually fold quite rapidly, in microseconds to minutes, depending on the protein and conditions. This is the so-called 'Levinthal Paradox'.

It is not really a paradox, of course. What it means is that polypeptides do not need to explore all possible conformations before they find the right one. Just as in any other rate

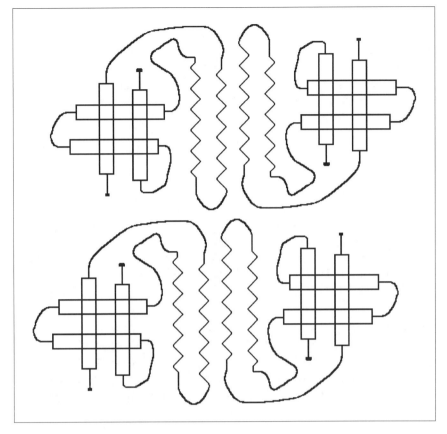

**Figure 1.6** Quaternary structure.

process, there are kinetic pathways or reaction mechanisms that direct the system to the required state, in the same way that the water molecules in a mountain stream do not need to try all possible paths before finding they should flow downhill.

However, what Levinthal was pointing out was that if we don't know these pathways for protein folding, a computational search for the correct fold—no matter how powerful our computers—is doomed to failure.

Repetition of the same $\phi$–$\psi$ angles from one amino acid to the next gives rise to a regular secondary structure element, of which $\alpha$-helix and $\beta$-sheets are the most common examples. In these structures the $\phi$–$\psi$ angles repeat in such a way that hydrogen bonds may form between different peptide groups to stabilize the structure.

The term 'random coil' is sometimes used incorrectly to designate non-regular structural elements within a protein structure. There is of course nothing random about this: the $\phi$–$\psi$ angles are well defined.

Many structural elements such as loops, turns or other motifs that determine the tertiary structure of the protein do not have a regular repeating $\phi$–$\psi$ signature, but are nonetheless unique.

One important feature is that, in samples of a particular protein (if pure and properly folded), all the molecules will have the same conformation, give or take a little bit of variation due to thermal fluctuation. This contrasts with the situation normally found in polymer chemistry, where the macromolecules rarely have a well-defined structure and samples are made up of a heterogeneous mix of conformations, quite often in dynamic interconversion.

A true 'random coil' is a hypothetical state in which the conformation ($\phi$–$\psi$ angles) of any one peptide group is totally uncorrelated with any other in the chain, and especially its neighbours.

Folded proteins are relatively unstable and can unfold ('denature') easily—especially with change in temperature or pH, or on addition of chemical denaturants such as urea, guanidine hydrochloride or alcohols. Denatured proteins have lost their tertiary and quaternary structure, but may retain some secondary structure features. They rarely approach the true random coil state.

Unfolded protein is also quite sticky stuff and has a tendency to aggregate with other denatured proteins or to stick to surfaces.

Traditional animal glues are made from denatured skin and bone. The main connective tissue protein, collagen, takes its name from the Greek word for glue.

This intrinsic stickiness of unfolded polypeptides appears to be one of the causes of prion diseases and other amyloid-related conditions such as mad cow disease, CJD and Alzheimer's. In such conditions, unfolded or misfolded proteins aggregate into lumps or 'plaques' that interfere with normal cell function.

## 1.3    Polynucleotides

The genetic information which encodes protein sequences is found in DNA (deoxyribonucleic acid), and the transcription and translation process involves RNA (ribonucleic acid). Both are polynucleotides consisting of long sequences of nucleic acids made up of a phosphoribose backbone with a choice of four different purine or pyrimidine side-chains or 'bases' attached (see Figures 1.7 and 1.8).

The specific, complementary base pairing in the double helical structures of DNA and RNA (Figure 1.9) is what gives rise to the ability to translate and proliferate this genetic information. (See Further Reading: *Biology for Chemists* for molecular biology details.)

When complementary strands of DNA and/or RNA come together, they form the characteristic right-handed double helix structures that lie at the heart of molecular biology. In the most common form ('B-DNA'), the base pairs stack in a twisted ladder-like conformation, with the purine–pyrimidine rings lying flat and perpendicular to the helix axis and spaced 0.34 nm apart. The negatively charged

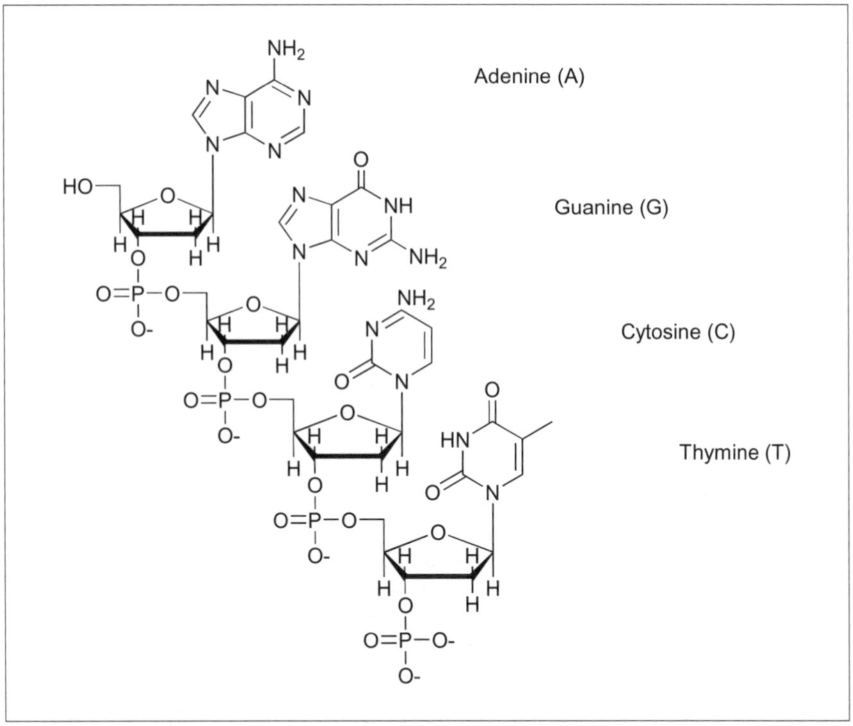

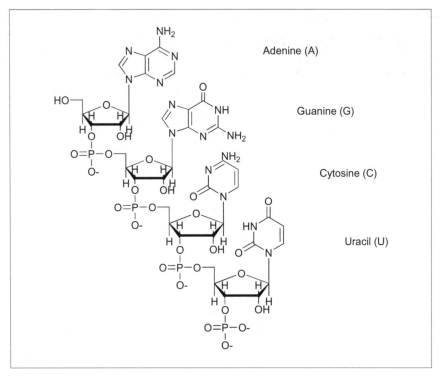

**Figure 1.7** DNA structure illustrating the deoxyribose–phosphate backbone to which may be attached purine (A, G) or pyrimidine (C, T) bases.

**Figure 1.8** RNA structure illustrating the sugar (ribose) - phosphate backbone, to which may be attached purine (A, G) or pyrimidine (C, U) bases.

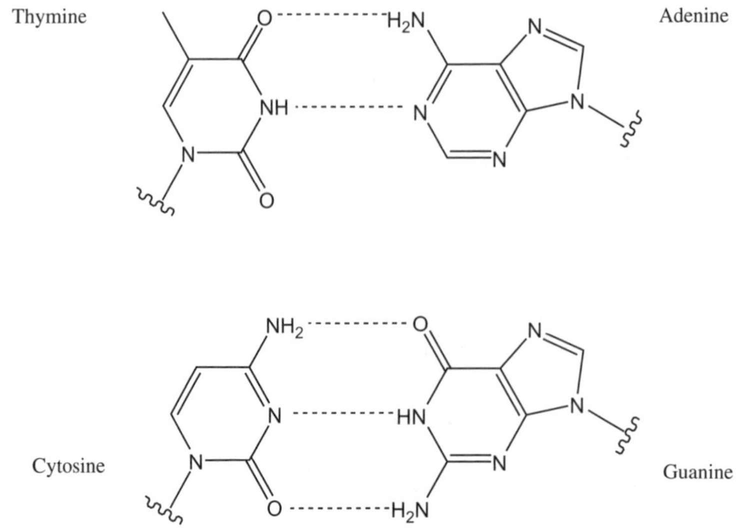

**Figure 1.9** Complementary base pairing (Watson–Crick) in DNA (RNA is similar, with uracil replacing thymine).

sugar–phosphate backbone lies to the outside of this cylindrical structure, which is roughly 2 nm in diameter.

One of the major challenges of current research is to understand how such long DNA polymers are packaged within the cell nucleus while still allowing access for genetic control and transcription. See Further Reading for more detail.

**Worked Problem 1.4**

**Q**: The DNA in each of your cells (*i.e.* the human genome) contains about three billion ($3 \times 10^9$) base pairs. How far would this stretch if laid out in a straight line?

**A**: Assume 0.34 nm spacing: $3 \times 10^9 \times 3.4 \times 10^{-10} = 1.02$ m.

Many other polynucleotide conformations are possible, including the left-handed helical 'Z-DNA' and more complicated structures thought to be involved in chain replication, together with super-coiling and more globular structures in single-stranded transfer RNA.

## 1.4 Polysaccharides

Complex polysaccharides such as starch, glycogen, cellulose, *etc.* play an important part in biochemistry both as energy stores and structural components. Many proteins are glycosylated ('glycoproteins'), with oligosaccharide chains (often branched) attached to specific amino acid residues, usually at the protein surface. The carbohydrate portion of glycoproteins is often involved in antigenicity, cell receptor and other molecular recognition processes.

Polysaccharides (and the smaller oligosaccharides) are polymers formed by linkage of individual sugar monomers, and may be linear (*e.g.* cellulose) or branched (*e.g.* glycogen).

Although some regular secondary structure is sometimes seen (*e.g.* in cellulose fibres), the complexity of chemical composition and polymer chain branching leads to much more disordered structures (or, at least, structures that are usually too complex to determine). This is why our understanding of polysaccharide structures and their interactions is still very poor.

Reminder: **glycosylation** is the covalent attachment of carbohydrate (sugar) groups; an **oligosaccharide** is a short chain polymer of sugars. Human blood groups are determined by the different oligosaccharides attached to glycoproteins and glycolipids on red blood cells.

## 1.5 Fats, Lipids and Detergents

Fats and lipids are common terms for those bits of biological organisms that are insoluble in water but can be extracted with organic solvents such as trichloromethane (chloroform), ethers, *etc.* They generally consist of a polar head group attached to non-polar tails of unbranched hydrocarbons. This amphiphilic nature (hydrophilic head, hydrophobic tail) gives this class of molecule important properties that are exploited both by biology itself, and by biophysical chemists in studying such systems.[2]

Broadly speaking, the number of hydrocarbon tails governs the behaviour in water.

**Detergents** generally contain a polar head group attached to a single non-polar tail (or equivalent). This allows them to form **micelles** in water, *i.e.* roughly globular assemblies of a number of molecules clustered together, with their head groups exposed to water, while their non-polar tails are buried inside the cluster and away from direct contact with the surrounding water (see Figure 1.10).

Detergents can solubilize or disperse other non-polar molecules in water. In the laboratory they can be used to solubilize membrane proteins. **Bile salts** are detergent-like molecules synthesized in the gall bladder and secreted in the small intestine to assist the dispersion and digestion of fats.

**Lipids** have two tails. This makes it difficult to pack the hydrocarbon chains effectively into a globular micelle structure, but they can form **lipid bilayers** instead (see Figure 1.11). Here the molecules form into two-dimensional arrays or sheets, in which two layers of lipids bury their tails inside, leaving the hydrophilic heads exposed either side to the water. These lipid bilayers provide the basic structures of cell membranes.

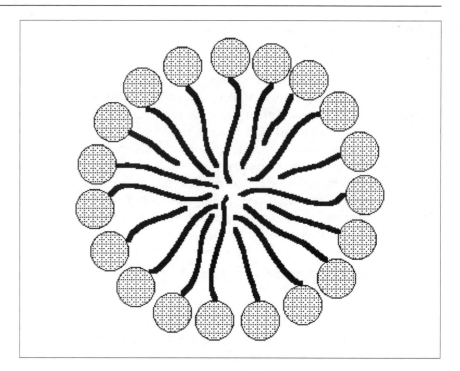

**Figure 1.10**  A micelle.

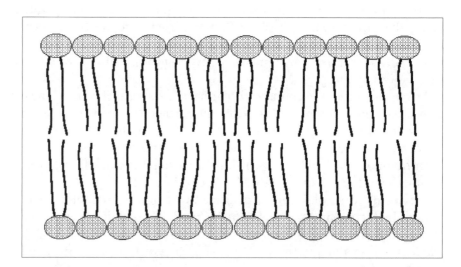

**Figure 1.11**  Lipid bilayer.

The **fluid mosaic model** pictures biological membranes as dynamic, two-dimensional seas of lipid bilayer, within which float the multitude of proteins and other molecules. These membrane-associated macro-molecules may be partially submerged in the lipid bilayer or may traverse the entire membrane. Other peripheral membrane proteins may be more loosely associated at the surface of the bilayer.

**Neutral fats** or **triglycerides** commonly have three tails (Figure 1.12). This makes it difficult to form a compromise between the hydrophilic head and the bulky hydrophobic tails, so these substances tend to be very insoluble and just form an amorphous mass in water. This is what we commonly see as 'fat'.

Triglycerides ('fats') act as concentrated, long-term metabolic energy stores (as opposed to glycogen, which can be metabolized more rapidly, but has a lower metabolic energy density).

The metabolic oxidation energy or 'calorific value' of carbohydrates (glycogen) is around $17 \, \mathrm{kJ \, g^{-1}}$ compared to about $39 \, \mathrm{kJ \, g^{-1}}$ for fats (triglycerides). Moreover polysaccharides absorb a lot of water (typically around 2 g water per gram of carbohydrate), so it takes almost 7 g of hydrated glycogen to provide the same energy as 1 g of fat. (This is why people on long-distance walks to the North or South Poles eat a high fat diet.)

Alkaline hydrolysis of the ester groups in triglycerides (*e.g.* conversion of glyceryl tristearate to sodium stearate) has been used since ancient times to convert fat into soap.

## 1.6    Water

Life evolved in an aqueous environment, and water is the major component in most biological organisms and tissues. Despite its

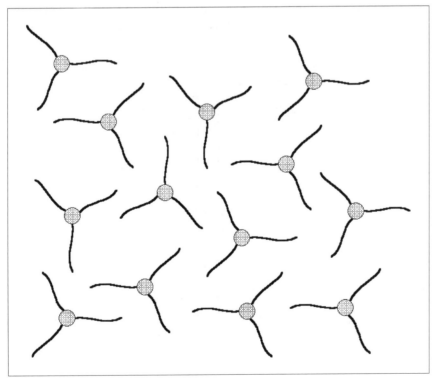

**Figure 1.12**   Fats.

The survival of life on Earth is often attributed to these anomalous volumetric properties. During winter, ice floats on the lake surface acting as a thermal insulator to prevent further freezing. Meanwhile, the bottom of the lake remains comfortably liquid at 4 °C.

familiarity, water is an unusual liquid in many ways, with several apparently anomalous properties.

Compared to molecules of a similar size, water has a much higher melting point and boiling point, and the liquid has an unusually high heat capacity, surface tension and dielectric constant.

Solid water (ice) at 0 °C has a lower density than the liquid, so ice floats on water. This volume contraction on melting continues as the temperature is increased to about 4 °C, where liquid water has its highest density under normal conditions (Figure 1.13).

All of these anomalous properties can be attributed to the polarity and hydrogen bonding ability of the water molecule. Because of the molecular structure, and the ability to act as both hydrogen bond donor and acceptor, the most favourable interactions occur in a tetrahedral arrangement in which one water molecule may interact with up to four neighbouring water molecules. Consequently, the normal structure of crystalline ice involves a quite open tetrahedral lattice linked by hydrogen bonding (Figure 1.14).

Most substances expand on heating because the increased thermal motion leads to larger average intermolecular distances. However when ice melts (at 0 °C), some of the hydrogen bonds break and the lattice becomes more flexible and dynamic, and some of the molecules can fall into the interstitial gaps to give a denser structure. This continues until around 4 °C where the gradually increasing thermal motion takes over, and the more usual thermal expansion occurs.

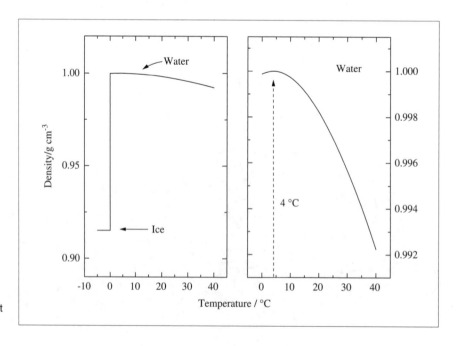

**Figure 1.13** The density of ice and liquid water as a function of temperature (at atmospheric pressure). Solid ice has a much lower density ($0.915\,g\,cm^{-3}$) than water (left panel). Liquid water (expanded scale, right panel) has a maximum density at around 4 °C.

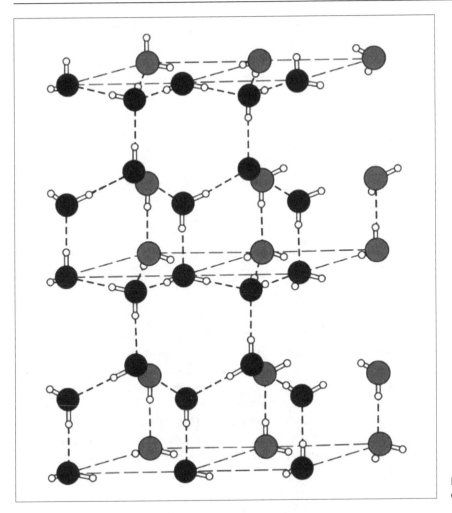

**Figure 1.14** The tetrahedral crystal structure of ordinary ice.

However, hydrogen bonding and residual tetrahedral structure persist in the liquid, although now in a much more dynamic and less ordered fashion, and to a lesser extent as the temperature rises. This residual hydrogen bonding contributes to the high heat capacity of liquid water. The heat capacity of a substance represents the energy required to raise the temperature of the substance by a given amount. In the case of liquid water, some of the energy goes into breaking intermolecular hydrogen bonds rather than molecular kinetic energy, so it takes more energy to bring about a rise in temperature than would otherwise be the case.

The high **surface tension** of water and its inability to wet greasy surfaces is also a consequence of this residual hydrogen-bonded structure. Water molecules at air–water interfaces or non-polar surfaces are less able to adopt the energetically preferred tetrahedral

'Temperature' is a measure of how we feel the kinetic energy of the random movements of atoms and molecules in any substance. The higgledy-piggledy rotational, vibrational, and translational motions of atoms and molecules increase with temperature.

arrangement. Soaps and detergents can overcome this by forming an amphiphilic layer at the interface.

The **dielectric constant** or **relative permittivity** ($\varepsilon_r$) of a substance is a measure of its polarizability in an electric field. For water at room temperature, $\varepsilon_r \approx 80$ (compared to one for a vacuum). This very high value arises because the dipolar water molecules tend to reorient and align parallel to the electric field. This has the effect of partially cancelling the electric field and results in a weakening of electrostatic interactions between charged groups. Figure 1.15

Remember that the electrostatic (Coulomb) potential energy between two charges, $q_1$ and $q_2$, separated by a distance $r$, is given by:

$$V_{qq} = q_1 q_2 / 4\pi\varepsilon_0\varepsilon_r r.$$

This is why salts are soluble and tend to dissociate into ions in water, but not in less polar solvents.

Consequently, the high value of $\varepsilon_r$ has a significant effect on interactions between charges in water.

### 1.6.1 Hydrophobicity

Because water molecules have such a high affinity for each other, non-polar molecules have difficulty fitting in to aqueous solutions. This is known as the **hydrophobic effect**. We know from experience that oil and water do not mix. Non-polar molecules are unable to form hydrogen bonds, so they cannot be accommodated easily within the partially hydrogen-bonded structures of liquid water. This leads to an apparent repulsion between water and non-polar molecules, such that the non-polar groups tend to be insoluble in the water and form separate phases or aggregates with other non-polar groups.

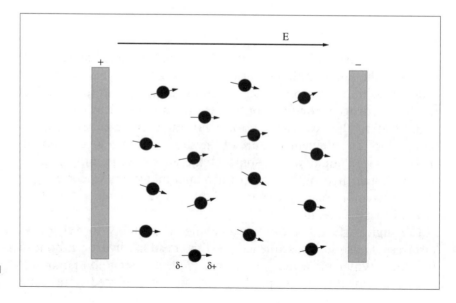

**Figure 1.15**  Molecular dipoles tend to align parallel to an electric field. Thermal motion will tend to disrupt this reorientation.

It is not that the individual water molecules repel the non-polar groups (they don't), but that the collective affinity of the water molecules for each other tends to exclude others that do not have similar hydrogen bonding tendencies.

One peculiar property of hydrophobic interactions is that they appear to get stronger with increasing temperature, at least at low temperatures. This is illustrated by the observation that the solubility in water of many non-polar compounds becomes smaller as the temperature is raised.[2]

Due to dipole : induced-dipole interactions, the attractive force between an isolated water molecule and a nearby non-polar molecule in a vacuum is probably greater than between two non-polar molecules under similar circumstances. That is why the build-up of static electricity tends to attract dust.

## 1.7    Foams, Surfactants and Emulsions

Although oil and water do not mix of their own accord, this can be done if we are willing to put work into it or use some means of circumventing the hydrophobic tendencies. Foams and emulsions are examples of dispersions of small air bubbles or fatty globules in water, and such materials play significant roles in food technology, cosmetics and related industries. They are normally produced by 'whipping' (mechanical agitation) or 'sparging' (blowing bubbles). It requires a lot of energy to overcome the water surface tension but this can be reduced by adding detergents or other surfactant molecules that reduce the effective surface tension by forming an amphiphilic layer at the interface—like a micelle (see Section 1.5), with the polar groups in contact with water and the non-polar tails towards the more hydrophobic air or oily globule.

This will be familiar from everyday experience with soaps, shampoos and other domestic cleaning products. But foams and emulsions are inherently unstable and will tend to collapse over time (see Box 1.2). Because of the high surface area to volume ratio, small bubbles or globules have a higher surface energy (more hydrophobic area exposed to water) and will merge to form larger bubbles or droplets. Denatured (unfolded) proteins often have good surfactant properties. Conversely, foaming will tend to denature proteins in solution and is best avoided.

Molecules with detergent-like properties are relatively rare in biology because of their tendency to disrupt biological membranes. However, there are some examples of proteins that have evolved to have biocompatible surfactant activities in specific instances.[3] Latherin is a protein from horse sweat that allows the sweat to spread better on the oily horsehair. Ranaspumins are proteins involved in constructing the foam nests of some tropical frogs. Hydrophobins are proteins produced by fungi to reduce surface tension and help them grow in thin water films. And our lungs contain specific phospholipid and carbohydrate-associated proteins that reduce surface tension forces between alveolar surfaces to help us breath more easily.

## Box 1.2 Surface Tension and the Pressure Inside a Bubble

Surface tension ($\gamma$) can be defined as the force per unit length at a line or edge in the liquid interface. ($\gamma = 0.073\,\mathrm{N\,m^{-1}}$ for water at an air interface.)

A bubble of radius $r$ will have an excess pressure, $\Delta P$, that can be estimated by considering the balance of forces indicated in Figure 1.16. At mechanical equilibrium, the force exerted by the

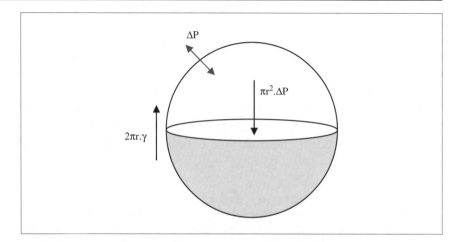

**Figure 1.16** The balance of forces in a bubble.

excess pressure on the circular cross-section area (area $\times$ pressure $= \pi r^2 \Delta P$) is balanced by the surface tension force in the perimeter (circumference $\times$ surface tension $= 2\pi r\gamma$).

$$\pi r^2 \Delta P = 2\pi r\gamma, \text{ giving } \Delta P = 2\gamma/r.$$

Consequently, the excess pressure inside a bubble is inversely proportional to its radius. Smaller bubbles will tend to coalesce or gas (air) will diffuse from one to the other to form larger bubbles. Solitary bubbles or bubbles at the surface of a foam will tend to burst if they can.

Similar principles apply to dispersions of fatty droplets in aqueous emulsions.

## 1.8    Acids, Bases, Buffers and Polyelectrolytes

The acid–base properties of water, together with its solvent polarity, mean that free charged groups (ions) are common and that most biological macromolecules must be regarded as polyelectrolyes, *i.e.* large molecules with multiple charges.

To summarize:

Water itself can dissociate : $\quad H_2O(l) \rightleftharpoons H^+(aq) + OH^-(aq)$

The bare proton ($H^+$ ion) does not really exists as a free entity in solution, but is better represented as $H_3O^+$, $[H_9O_4]^+$, or other complex species. However, $H^+$ is a convenient shorthand.

The equilibrium constant for this (remembering the thermodynamic convention that the activity of $H_2O(l) = 1$) is $K_W = [H^+][OH^-] = 10^{-14} \, mol^2 \, dm^{-6}$ at 25 °C.

For hypothetically pure water at $25\,^{\circ}C$, $[H^+]=[OH^-]=10^{-7}\,mol\,dm^{-3}$.

Hydrogen ion concentration is more conveniently expressed using the logarithmic pH scale, in which $pH = -\log_{10}[H^+]$.

Strictly speaking, we should talk in terms of 'activity' rather than 'concentration' here. The thermodynamic activity of a solute is its concentration multiplied by a fudge factor (called 'activity coefficient') that takes account of some of the intermolecular interactions in solution. For dilute solutions, the difference is rarely significant.

---

### Worked Problem 1.5

**Q**: The measured pH of 'ultrapure' laboratory water is frequently below pH 7. Why might this be?

**A**: Several reasons:

 a. Dissolved atmospheric $CO_2$ (carbonic acid) if the water has been left standing for a while

 b. pH not measured at $25\,^{\circ}C$ ($H^+$ dissociation increases with temperature)

 c. Contamination from an unwashed pH electrode

 d. pH meter wrongly calibrated

---

Acidic and basic groups in solution can take part in this equilibrium exchange of protons:

$$AH \rightleftharpoons A^- + H^+$$

with the **acid dissociation constant**, $K_A = [A^-][H^+]/[AH]$ and $pK_A = -\log_{10}K_A$.

The $pK_A$ of any group is most conveniently viewed as that pH at which the conjugate acid is 50% dissociated ($[A^-]=[AH]$, so that $K_A=[H^+]$ in these circumstances).

In proteins, the relevant groups are the acidic and basic amino acid side chains, and the $N$- and $C$-terminal peptide groups (Table 1.2). As a consequence, the overall charge on a protein molecule will depend on pH.

Interactions with other groups and change in solvent environment in folded proteins can affect actual pK values.

---

### Worked Problem 1.6

**Q**: Lysozyme is a small globular protein with antibiotic activity found in a variety of biological fluids. Typically it consists of a single polypeptide chain of around 129 amino acids (RMM 14 300), containing two glutamic acid (Glu), seven aspartic acid (Asp), six lysine (Lys), 11 arginine (Arg), three tyrosine (Tyr) and one histidine (His) residues, in addition to numerous other groups. What total charge might this protein have at pH 2, pH 7 or pH 12?

**Table 1.2**  Typical $pK_A$ and charge state for protein amino acid residues and other groups in water (see Figure 1.2 for amino acid structures and abbreviations)

| Group | $pH < pK_A$ | Typical $pK_A$ | $pH > pK_A$ |
|---|---|---|---|
| C-terminus | –COOH | 3 | -COO$^-$ |
| Glu, Asp | –COOH | 4 | -COO$^-$ |
| His | –Im–H$^+$ | 6 | –Im |
| N-terminus | –NH$_3^+$ | 8 | –NH$_2$ |
| Cys | –SH | 8 | –S$^-$ |
| Lys | –NH$_3^+$ | 11 | –NH$_2$ |
| Tyr | –$\phi$–OH | 11 | –$\phi$–O$^-$ |
| Arg | –C(NH$_2$)$_2^+$ | 12.5 | –C(NH)(NH$_2$) |
| Phosphoglycerol | R–P(OH)O$_2^-$ | 5.6 | R–PO$_3^{2-}$ |
| R = CH$_2$(OH)CH(OH)CH$_2$-O | | | |

**A**: At pH 2 (below the typical $pK_A$ of all the protein groups), Arg, Lys, His and N-terminal groups are positively charged, all others are neutral. Consequently the total charge on the protein is: $6(\text{Lys}) + 11(\text{Arg}) + 1(\text{His}) + 1(\text{N-terminus}) = +19$.

At pH 7 (pH > $pK_A$ for Asp, Glu, His and C-terminal groups, but pH < $pK_A$ for others), total charge = $-2(\text{Glu})–7(\text{Asp})–1(\text{C-terminal}) + 6(\text{Lys}) + 11(\text{Arg}) + 1(\text{N-terminus}) = +8$.

At pH 12 (pH > $pK_A$ for all groups other than Arg), total charge = $-2(\text{Glu}) - 7(\text{Asp}) - 1(\text{C-terminal}) - 3(\text{Tyr}) + 11(\text{Arg}) = -2$. Note that these are only approximate estimates, since actual $pK_a$ values will depend on local environments within the protein.

Typical pH values for biological and other fluids are as follows:

- blood pH 7.3–7.5
- gastric juices pH 1–3
- saliva pH 6.5–7.5
- urine pH 5–8
- milk pH 6.3–6.7
- beer pH 4–5
- wine pH 2–4
- soft drinks pH 2–4
- citrus fruits pH 1.8–4.

At low pH, most proteins will carry a net positive charge, whereas their overall charge will be negative at very high pH. The intermediate pH at which the net charge is zero is called the **isoelectric point**, pI. At this pH, the numbers of positively and negatively charged groups just balance. This will depend on the actual composition and conformation of the protein, and will determine its behaviour (mobility) in an electric field or in certain chromatographic processes (see Chapter 7) as well as its functional properties.

DNA and RNA will carry a net negative charge at neutral pH because of the phosphate backbone. (The purine and pyrimidine bases themselves are uncharged.)

Many lipids and carbohydrates may contain acidic or basic groups, so their charges may also depend on pH.

Because of the way in which molecular charge can affect biological properties, it is usually necessary to control the pH. Buffer solutions contain mixtures of conjugate acids and bases that can tolerate addition of $H^+$ or $OH^-$ without great change in pH. Buffering power is greatest when the desired pH is close to the $pK_A$ of the buffer components.

Aqueous buffer solutions are frequently made up by adding similar amounts of a weak acid and its salt with a strong base (*e.g.* ethanoic acid and sodium ethanoate) to water (Table 1.3). Alternatively, one may start with a solution of weak acid (say) and titrate to the desired pH by addition of base.

Oxygen binding by haemoglobin is very sensitive to pH. This is one of the reasons why pH control in the bloodstream and other tissues is particularly important.

The equivalent for basic buffers is to mix the weak base and its salt with a strong acid (*e.g.* ethylamine and ethylamine-HCl), or to titrate the weak base solution with strong acid.

---

### Box 1.3 How to Estimate Buffer pH

Buffering in aqueous solution is based on a weak acid–base equilibrium:

$$AH \rightleftharpoons A^- + H^+; \qquad K_A = [A^-][H^+]/[AH]$$
$$\text{so that} \qquad [H^+] = K_A[AH]/[A^-].$$
$$\text{If}[AH]/[A^-] = 10, \ \text{then}[H^+] = 10 \times K_A \ \text{and} \ pH = pK_A - 1.$$
$$\text{If}[AH]/[A^-] = 1, \ \text{then}[H^+] = K_A \ \text{and} \ pH = pK_A.$$
$$\text{If}[AH]/[A^-] = 0.1, \ \text{then}[H^+] = 0.1 \times K_A \ \text{and} \ pH = pK_A + 1.$$
$$[\text{Reminder}: p(\text{something}) = -\log_{10}(\text{something})]$$

Buffering capacity is best around $pH = pK_A$ since there is likely to be sufficient [AH] and [A$^-$] in solution to accommodate addition of small amounts of strong base or acid without the pH straying outside this range.

---

## 1.9    A Note about Units

Scientists are (mostly) human. Although we try to be systematic, logical, and consistent, we do have occasional lapses—sometimes for sheer laziness, sometimes for practical convenience. Consequently, and especially in an interdisciplinary subject such as this, you will encounter non-standard units and different terminologies. Although you will be familiar with SI units (and this book tries to be consistent in their use), you will already know that they are not always used elsewhere. For example, despite metrication, we still buy petrol in litres (or even gallons) and not cubic decimetres ($dm^3$).

**Table 1.3**  Typical buffers and their useful pH ranges

| Buffer | $pK_A$ | AH | $A^-$ | pH range | | |
|---|---|---|---|---|---|---|
| | | | | $[AH]/$ $[A^-]=10$ | $[AH]/$ $[A^-]=1$ | $[AH]/$ $[A^-]=0.1$ |
| Ethanoic acid/ sodium ethanoate | 4.8 | $CH_3COOH$ | $CH_3COO^-$ | 3.8 | 4.8 | 5.8 |
| Carbonic acid/ sodium carbonate | 6.4 | $H_2CO_3$ | $HCO_3^-$ | 5.4 | 6.4 | 7.4 |
| $NaH_2PO_4$/ $Na_2HPO_4$ | 7.2 | $H_2PO_4^-$ | $HPO_4^{2-}$ | 6.2 | 7.7 | 8.2 |
| Ethylamine/ ethylamine-HCl | 9 | $C_2H_5NH_3^+$ | $C_2H_5NH_2$ | 8 | 9 | 10 |
| | | | | lowest useful pH | best buffering pH | highest useful pH |

Here is a (partial) list of commonly used units and terms, together with their more consistent (SI) equivalents:

| | Non-standard | Standard (SI) |
|---|---|---|
| Length | micron ($\mu$) | $10^{-6}\,m$ |
| | Angstrom unit, Å | $10^{-10}\,m$ |
| Volume | litre, L | $dm^3$ |
| | mL | $cm^3$ |
| | $\mu$L, microlitre | $\mu dm^3$ |
| Concentration | mg per mL, $mg\,mL^{-1}$ | $mg\,cm^{-3}$ |
| | M, molar, molarity, moles per litre | $mol\,dm^{-3}$ |
| Relative molecular mass | 'molecular weight' | RMM |
| | 1 Dalton (1 Da) | 1 amu |
| Heat energy | calorie (cal) | Joule (J) |
| | 1 cal | 4.184 J |

## Summary of Key Points

1. Biological systems are made up of structured macromolecules of specific sequence (proteins, nucleic acids, polysaccharides) and of smaller molecules (lipids, *etc.*) that self-assemble into larger structures.
2. Secondary, tertiary, quaternary structures, assembly and interactions involve non-covalent forces.
3. Water plays a dominant role in these interactions.

## Problems

**1.1.** Serum albumin (RMM approx. 65 000) is present in blood at a concentration of around $45\,mg\,cm^{-3}$. Roughly how far apart are the protein molecules and how might this compare to their size?

**1.2.** (a) How many possible conformers are there for a polypeptide made up of 100 amino acids? (b) Assuming that these conformers may switch on a femtosecond timescale (about the fastest possible for bond rotations), how long might it take to explore all possible conformers?

**1.3.** How high might an average 70 kg person climb (or jump) on the energy provided by (a) 10 g of sugar; (b) 10 g of fat? Is this realistic? [Reminder: 'calorific values' are around $39\,kJ\,g^{-1}$ for fat and $17\,kJ\,g^{-1}$ for carbohydrate.]

**1.4.** The average resting male produces about 7000 kJ of heat energy per day. How much fat might he therefore lose by just sitting around and doing nothing? [Reminder: the 'calorific value' of fat is around $39\,kJ\,g^{-1}$.]

**1.5.** Why is the water temperature at the bottom of a (partially) frozen lake usually around $4\,°C$?

**1.6.** (a) List some of the anomalous properties of water that can be ascribed to hydrogen bonding. (b) Is this the bond that sank the *Titanic*?

**1.7.** What is the electrostatic potential energy between a pair of $Na^+$ and $Cl^-$ ions, 0.5 nm ($5\,\text{Å}$) apart: (a) in vacuum; (b) in water?

**1.8.** (a) Why does the relative permittivity (dielectric constant) of water decrease with temperature? (b) How might this affect the strength of electrostatic interactions between opposite charges in water? (c) Are such interactions endothermic or exothermic?

**1.9.** Popular newspapers sometimes state that the DNA molecule in a human cell is about 1.8 metres long or the height of Craig Venter (one of the scientists who led the Human Genome Project). Why might this differ from the estimate given in Worked Problem 1.4?

## References

1. C. Levinthal, Are there pathways for protein folding?, *J. Chim. Phys.*, 1968, **65**, 44–45.
2. C. Tanford, *The Hydrophobic Effect: Formation of Micelles and Biological Membranes*, Wiley Interscience, New York, 1973.
3. A. Cooper and M. W. Kennedy, Biofoams and natural protein surfactants, *Biophys. Chem.*, 2010, **151**, 96–104.

## Further Reading

J. M. Berg, J. L. Tymoczko and L. Stryer, *Biochemistry*, Freeman, San Francisco, 6th edn, 2006.

S. Doonan, *Peptides and Proteins*, RSC Tutorial Chemistry Text, Royal Society of Chemistry, Cambridge, 2002.

S. Mitchell and P. Carmichael, *Biology for Chemists*, RSC Tutorial Chemistry Text, Royal Society of Chemistry, Cambridge, 2004.

N. C. Price, R. A. Dwek, R. G. Ratcliffe and M. R. Wormald, *Physical Chemistry for Biochemists*, Oxford University Press, Oxford, 3rd edn, 2001.

D. Sheehan, *Physical Biochemistry: Principles and Applications*, Wiley, New York, 2nd edn, 2009.

K. E. van Holde, W. C. Johnson and P. S. Ho, *Principles of Physical Biochemistry*, Prentice Hall, New York, 1998.

# 2
# Spectroscopy

The interaction of light with matter is the way most of us see the world around us. Spectroscopic techniques provide some of the more powerful experimental methods for determining the structure and properties of molecules—not just biological ones.

## Aims

- In this chapter we revise the basic properties of electromagnetic radiation and explore experimental spectroscopic methods for studying biomolecules in solution. By the end you should be able to:
- Explain the nature of electromagnetic radiation and its interaction with matter
- Describe the basic mechanisms of absorbance, fluorescence, circular dichroism, Raman, magnetic resonance and related spectroscopies
- Describe instrumental methods for each of these techniques
- Explain how different factors affect the observed spectral properties of biomolecules
- Apply this understanding to the interpretation of biomolecular properties

## 2.1 Electromagnetic Waves and their Interactions

Electromagnetic waves are produced whenever a moving electric charge alters speed or direction. This is a consequence of the laws of electromagnetism formulated by Michael Faraday and James Clerk Maxwell in the 19th century and first demonstrated experimentally by Hertz. The same principles still apply today, though sometimes modified in detail by quantum effects.

The most familiar everyday example is a radio transmitter aerial (or antenna). In its simplest form, this is just a length of wire acting as conductor for an oscillating electric current (Figure 2.1). As the electrons

Around 1886, the physicist Heinrich Hertz was experimenting with high voltage spark discharges from an induction coil. He noticed that these oscillatory discharges induced small sparks in metal objects some distance away. This was the first observation of the wave-like transmission of electromagnetic energy first predicted theoretically by James Clerk Maxwell in 1864. This soon led to Marconi's developments of radio communication.

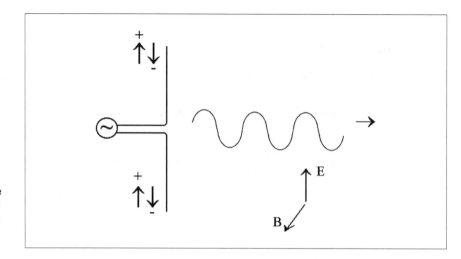

**Figure 2.1** Sketch of a simple radio antenna ('aerial') in which an oscillating charge or electric current produces a radiating electromagnetic field.

The heating effect in a microwave oven is mainly due to induction of molecular rotations in water molecules in the sample. The rotating water molecules quickly collide with neighbouring molecules to generate the more random molecular motions that we know as heat. Most other non-conducting materials are transparent to microwaves at this frequency, so the heating effect can be distributed throughout.

oscillate up and down the wire, they produce an oscillating electric and magnetic field (at right angles to each other) that propagates outwards as an electromagnetic wave. The frequency in this case is determined by the resonant frequency of the electrical circuitry that drives the oscillations.

Absorption of electromagnetic radiation by the receiver is just the opposite of this process. When the oscillating electromagnetic wave encounters an appropriate antenna, an oscillating charge (electric current) and/or magnetic dipole is induced in the antenna. This absorption of electromagnetic energy is most effective if the antenna is 'tuned' to match the frequency of the incoming radiation (or some multiple 'overtone' of that frequency).

This analogy may be applied on the molecular scale as well. For example, in the absorption of ultraviolet (UV) and visible (vis) radiation by atoms and molecules, the 'antennae' may be visualized as the outermost electrons hopping between orbitals (Figure 2.2). For infrared (IR), it is usually the bond vibrations of dipolar groups that give rise to the effects and rotations of such groups (or polar molecules) can be related to absorption of microwave radiation. At the nuclear level, reorientations of nuclear magnetic dipole moments of certain atomic nuclei (acting like tiny bar magnets or compass needles) can also act as antennae for electromagnetic radiation—usually in the radio frequency or microwave range.

All electromagnetic radiation, regardless of its source, travels at the same velocity ('the speed of light') in a vacuum. However, the electromagnetic spectrum (Figure 2.3) covers a wide range of frequencies and wavelengths connected by the equation:

$$c = f\lambda$$

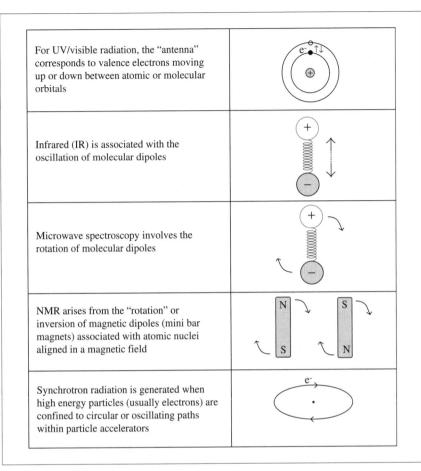

| | |
|---|---|
| For UV/visible radiation, the "antenna" corresponds to valence electrons moving up or down between atomic or molecular orbitals | |
| Infrared (IR) is associated with the oscillation of molecular dipoles | |
| Microwave spectroscopy involves the rotation of molecular dipoles | |
| NMR arises from the "rotation" or inversion of magnetic dipoles (mini bar magnets) associated with atomic nuclei aligned in a magnetic field | |
| Synchrotron radiation is generated when high energy particles (usually electrons) are confined to circular or oscillating paths within particle accelerators | |

**Figure 2.2** Cartoons of molecular 'antennae'—the movement of charges responsible for emission and absorption of electromagnetic radiation at the molecular level.

where $f$ is the frequency [in Hertz (Hz) or s$^{-1}$], $\lambda$ is the wavelength (in metres) and $c$ is the velocity of light ($= 3 \times 10^8\,\mathrm{m\,s^{-1}}$ in a vacuum).

For convenience, wavelengths are also often quoted in other units such as nanometres (nm), microns (1 micron = 1 micrometre = $10^{-6}$ m) or Angstrom units (Å, 1 Å = $10^{-10}$ m).

It is also customary (in vibrational spectroscopy) to refer to the reciprocal wavelength or wavenumber ($1/\lambda$) as the 'frequency' (in cm$^{-1}$).

The velocity of light in transparent materials (*e.g.* water, glass) is less than in the vacuum, and this difference in velocity is measured by the 'refractive index' of the material:

Refractive index, $n$ = speed of light in the material/$c$

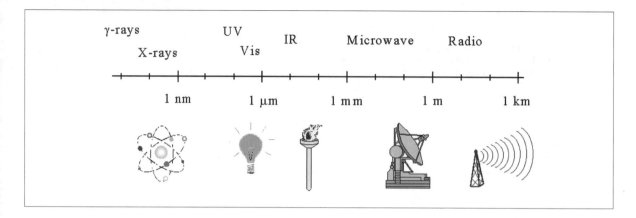

**Figure 2.3** The electromagnetic spectrum.

Differences in refractive index in solution can be used to detect, for example, concentration gradients or moving boundaries of macromolecules in ultracentrifuges (see Chapter 4).

This gives rise to the phenomenon of 'refraction', where a beam of light may be bent when it encounters a change in refractive index. In general, $n$ depends on the wavelength or frequency of the electromagnetic radiation. This gives rise to familiar 'dispersion' effects in lenses or prisms, where a beam of light may be split into its component wavelengths or 'spectrum'.

Electromagnetic waves have all the properties of waves in general: reflection, refraction, interference, diffraction. However, through the magic of quantum theory ('wave–particle duality'), they may also behave like quantized particles or 'photons'. The energy ($E$) of any photon is related to its frequency:

The photoelectric effect is the emission of electrons from metal surfaces exposed to light. Surprisingly, it is the wavelength of the light (not its intensity) that determines whether an electron is ejected: below a given wavelength (characteristic of the metal), no electrons will be ejected no matter how strong the light source. This effect was explained in terms of quantum theory by Albert Einstein in 1905. He subsequently received the Nobel Prize for this work (rather than his theories of relativity, started in the same year).

$$E = \mathrm{h}f = \mathrm{h}c/\lambda$$

where h is the universal Planck constant (h $= 6.626 \times 10^{-34}$ Js). These quantum effects give rise to observations such as the photoelectric effect, which is impossible to understand by wave theory alone.

Most of the time, this wave–particle duality allows us to use either photons or waves to describe phenomena—whichever is more convenient or easier to visualize.

### Worked Problem 2.1

**Q**: The human retina is most sensitive to light at around 500 nm. What is the energy of a single photon of this wavelength? What does this correspond to in kJ mol$^{-1}$? How does this compare to typical bond energies?

**A**: For one photon: $E = hf = hc/\lambda = 6.626 \times 10^{-34} \times 3 \times 10^{8}/500 \times 10^{-9} = 4 \times 10^{-19}$ J.

Multiply by $N_A$ (Avogadro's number) to give, for one mole of photons, $4 \times 10^{-19} \times 6 \times 10^{23}/1000 = 240$ kJ mol$^{-1}$.

This is somewhat below the energy typically required to break a covalent bond (*e.g.* C–C bond energy is around 340 kJ mol$^{-1}$).

Visible light is absorbed in the retina by 11-*cis* retinal, a conjugated polyene chromophore forming part of the rhodopsin visual pigment protein. This results in efficient photo-isomerization of the chromophore to the all-*trans* form. The role of the protein is to control the wavelength at which the retinal absorbs, to lower the energy barrier to specific *cis–trans* isomerization in the electronically excited state, and to trigger subsequent biochemical processes that lead to nerve impulses sent to the brain.

### 2.1.1 Absorbance: The Beer–Lambert law

The extent to which electromagnetic radiation is absorbed by a sample is expressed in terms of the absorbance, $A$, or transmittance, $T$.

If $I_0$ is the intensity of the incident radiation and $I$ is the intensity transmitted, then the transmittance (usually expressed as a percentage) and absorbance are given by:

$$T(\%) = 100 \times I/I_0$$

$$A = -\log_{10}(I/I_0) = -\log(T)$$

The absorbance, $A$, is a particularly useful quantity since it is directly proportional to the optical density or thickness of the sample. The reason for this is shown in Figure 2.4.

Imagine the object split in two (Figure 2.4), each with a transmittance (say) of 20%, for example. The absorbance of each half would therefore be $-\log(0.2) = 0.699$. Of the 20% transmitted by the first half ($I_1$), only a further 20% gets through the second half, giving a final transmittance for the whole sample of just 4%. The absorbance of the whole sample, on the other hand is $-\log(0.04) = 1.398 = 2 \times 0.699$, or just the sum of the two separate absorbances.

By imagining the sample sliced into arbitrarily small sections (Figure 2.5), it can be shown that the light intensity falls off exponentially with distance through a homogeneous substance.

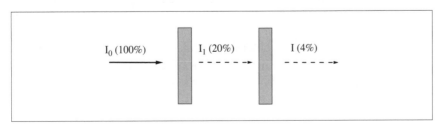

$I_0$ (100%)    $I_1$ (20%)    $I$ (4%)

**Figure 2.4** Sequential absorption of light by separate segments.

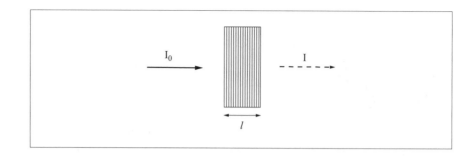

**Figure 2.5** Sequential absorption of light arbitrarily small section.

In the particular case of uniformly absorbing samples such as solutions or gases, this leads to the Beer–Lambert Law:

$$A = \varepsilon c l$$

The absorbance (*A*) of a sample is a dimensionless number, whereas the molar extinction coefficient (ε) has the units $mol^{-1} dm^3 cm^{-1}$. For solutions, ε can be visualized as the absorbance at a particular wavelength for a (hypothetical) sample of 1 molar concentration ($1\,mol\,dm^{-3}$) measured with a 1 cm path length.

where ε is the (molar) extinction coefficient, *c* is the (molar) concentration of the absorbing species, and *l* is the path length or thickness of the sample (usually in cm). The extinction coefficient (ε) is a number characteristic of the absorbing molecules at the wavelength of the incident radiation. It can also depend on the environment in which the molecules are located, *e.g.* solvent polarity (see Section 2.2.5). The advantage of using absorbance (*A*) is that, unlike transmittance (*T*), it is directly proportional to the concentration and path length of the sample.

### 2.1.2 Absorption Cross-section

It is sometimes useful to picture the extinction coefficient, ε, as the (hypothetical) cross-section area of the chromophore that is absorbing the light. The **absorption (or absorbance) cross-section** (σ) is numerically related to ε as follows:

$$\sigma = 3.8 \times 10^{-21}\, \varepsilon\, cm^2$$

---

**Worked Problem 2.2**

**Q**: Prove that $\sigma = 3.8 \times 10^{-21}\, \varepsilon\, cm^2$ (requires some competence with calculus).

**A**: Consider a volume element in the solution of 1 cm cross-sectional area and length d*l* (in cm). The number of molecules (chromophores) in this volume element will be $10^{-3}\, N_A c.dl$, where *c* is the chromophore concentration (in $mol\,dm^{-3}$). If each of these chromophores has an absorbance cross-section, $\sigma$, then the fraction of light absorbed by this volume element is:

$$dI/I = -\sigma\, 10^{-3}\, N_A c.dl$$

Integration over a finite path length, $l$, gives

$$ln(I/I_0) = -\sigma\, 10^{-3}\, N_A cl$$

However, from the definition of absorbance and the Beer–Lambert law:

$$A = -\log_{10}(I/I_0) = -ln(I/I_0)/2.303 = \sigma 10^{-3} N_A cl/2.303 = \varepsilon cl$$

Hence $\sigma = 2.303\, \varepsilon/10^{-3}\, N_A = 3.8 \times 10^{-21}\, \varepsilon\, cm^2$.

---

**Worked Problem 2.3**

**Q**: What is the absorption cross-section, $\sigma$, at 280 nm for the tryptophan chromophore whose molar extinction coefficient $\varepsilon_{280} = 5600\ (mol\,dm^{-3})^{-1}\,cm^{-1}$? How does this compare to the actual physical dimensions of this molecule?

**A**: $\sigma = 3.8 \times 10^{-21} \times 5600 = 2.1 \times 10^{-17}\, cm^2$

This corresponds to a square of side $\approx 5$ nm (0.5 Å), which is much less than the actual molecular dimensions.

## 2.1.3 Turbidity and Light Scatter

Often the passage of radiation ('light') through material is reduced not by absorbance but by scattering of the radiation by suspended particles or other inhomogeneities (**Rayleigh scattering**). For example, very little light passes directly through a glass of milk because of (multiple) scattering off the fine protein/fat globules in the milk. None of the components in the milk actually absorb much light in the visible region, but they are in the appropriate size range to scatter visible light—that is why the milk 'looks' white in colour: what we see is the light scattered back to our eyes.

The mathematical treatment of absorbance in such turbid samples is quite complex, and the simple Beer–Lambert law does not necessarily apply, except as a useful approximation in certain cases.

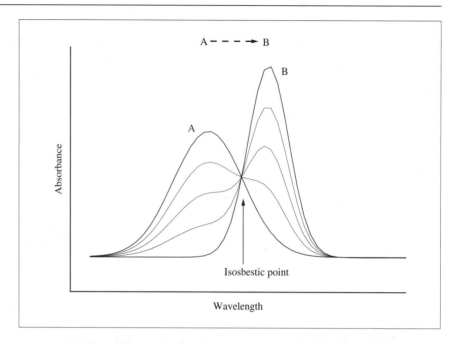

**Figure 2.6** Superimposed spectra showing an isosbestic point.

**Dynamic light scattering** (Section 4.6) can be used to give information about the size and heterogeneity of small particles and macromolecules in solution or suspension.

### 2.1.4  Isosbestic Points

Often a sample may be made up of a mixture of two species in equilibrium with each other, with different but overlapping absorbance spectra. In such cases there will be one (or more) wavelength(s) at which the absorbances of the two species are the same. This is known as an **isosbestic point** (Figure 2.6). If the equilibrium shifts (as in the $A \rightleftharpoons B$ example in Figure 2.6), the superimposed spectra will all cross at this point. The existence of an isosbestic point is a good indicator of a true equilibrium process and the absorbance at this wavelength can be a useful experimental reference point.

## Box 2.1 Proof of existence of an isosbestic point for a 2-state equilibrium process

Consider a mixture of compounds (A, B) in stoichiometric equilibrium with each other:

$$A \rightleftharpoons B$$

The absorbance of this mixture at any wavelength, $A(\lambda)$, will depend on the concentrations of each of the substances and their molar extinction coefficients at that wavelength:

$$A(\lambda) = \varepsilon_A(\lambda).[A] + \varepsilon_B(\lambda).[B]$$

If there happens to be some wavelength, $\lambda_{iso}$, at which both the molar extinction coefficients are the same [*i.e.* where the two spectra overlap, $\varepsilon_A(\lambda_{iso}) = \varepsilon_B(\lambda_{iso})$], then the absorbance at this wavelength will be fixed:

$$A(\lambda_{iso}) = \varepsilon(\lambda_{iso}).([A] + [B]) = \text{constant}$$

This is because the total concentration ($[A]+[B]$) remains constant at equilibrium regardless of the composition.

Note that more complex, non-two-state equilibria involving significant accumulation of intermediate species (*e.g.* A $\rightleftharpoons$ X $\rightleftharpoons$ B) will not usually show an isosbestic point since it is unlikely that all the spectra (A, X, B, *etc.*) cross at the same point.

## 2.2    UV/Visible Spectroscopy

Absorption of light in the visible or ultraviolet regions of the spectrum is usually due to electronic transitions in the molecules or substituent groups. For example, the familiar red colour of blood is due to absorption of visible light around 410 nm by the haem group in the oxygen transport protein, haemoglobin. Plants appear green because of long wavelength red light absorption by chlorophyll and other photosynthetic pigments. Indeed, the mechanism of vision itself relies on photon absorption by photosensitive pigments ('rhodopsins') in the photoreceptor cells of the retina at the back of the eye.

Colour is not always due to absorbance of light. For example, the brilliant hues seen in butterfly wings or bird feathers is often due to diffraction or interference effects arising from structural regularities—as in diffraction gratings or oil films.

Light in the near-UV region (around 250–300 nm) is absorbed by aromatic side chains in proteins and by the purine and pyrimidine bases of nucleic acids, whilst most chemical groups will begin to absorb in the far UV ($<250$ nm). We look in more detail at the characteristic spectra of these groups later. Here we consider how to measure these experimentally.

### 2.2.1    Measurement of Absorbance

#### 2.2.1.1    Spectrophotometers

Spectrophotometers for the measurement of absorbance in the UV/visible range come in a variety of configurations. The most common

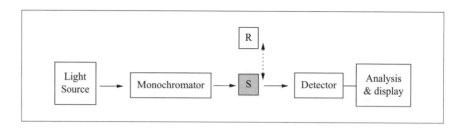

**Figure 2.7** Single beam spectrophotometer where S is the sample and R is the reference.

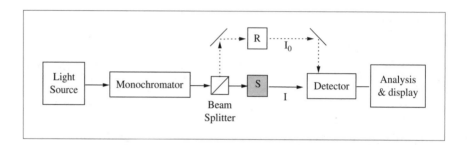

**Figure 2.8** Double beam spectrophotometer.

routine laboratory instruments are single- or double-beam devices made up of a light source, monochromator, sample compartment, detector, data processor and display (Figures 2.7 and 2.8).

The light source is commonly a tungsten filament lamp (covering the visible region of the spectrum, typically 340–800 nm) or a deuterium discharge lamp for the UV region (200–350 nm). For some applications, more intense sources such as a xenon or xenon/mercury arc lamp, or even lasers, may be required.

With the exception of lasers, light from such sources covers a wide spectral band, and some wavelength selection is required. The light from the source is focused through the monochromator – usually a diffraction grating, or simple colour filters in more basic instruments – and on to the sample or reference. In single beam instruments, the sample and reference solutions must be switched manually in or out of the beam and measured for each wavelength separately. In double-beam devices the light beam is split so that sample and reference solutions may be monitored simultaneously. Light then enters the detector so that the intensity passing through the sample ($I$) and through the reference ($I_0$) may be compared and analysed electronically to give T or A, as required.

Liquid samples (solutions) are normally contained in a rectangular quartz (UV) or optical glass (visible) cell (called a 'cuvette'), with transparent faces and defined path length. Disposable plastic cuvettes are also available for more routine work. The reference cuvette is used to measure the light transmitted in the absence of sample ($I_0$). It should, therefore, match the optical properties of the sample cuvette, and would normally contain just the solvent or other appropriate control.

Very small sample volumes (0.5–2 µl) can also now be analysed in a newer generation of instruments that do not require a cuvette. These are essentially single-beam spectrophotometers in which the tiny sample droplet is retained by capillary action between the faces of fibre-optic light guides; one for illumination and one leading to the detector. The path length of the entrapped sample is typically very small (0.05–1 mm). The light source is usually a xenon flash lamp with CCD array detection (see below, and Figure 2.10, for example).

For routine assays requiring analysis of large numbers of samples, 'micro-titre' or 'multi-well plate readers' have been developed. These consist of single-beam instruments in which the samples are contained in standard 96-well (8 × 12) plates, each well containing a separate sample. This allows large numbers of samples to be measured rapidly, often automatically under robotic control. The multi-well/micro-titre plates themselves are often coated with specific reagents so that particular analyses may be done rapidly and routinely *in situ*.

### 2.2.1.2 Photodetectors: Photomultipliers, Diode Array and CCDs

The photodetector is central to any photometric method and a number of techniques are used.

A **photomultiplier** (Figure 2.9) is an electronic device for detecting photons based on the photoelectric effect. It is made up of a series

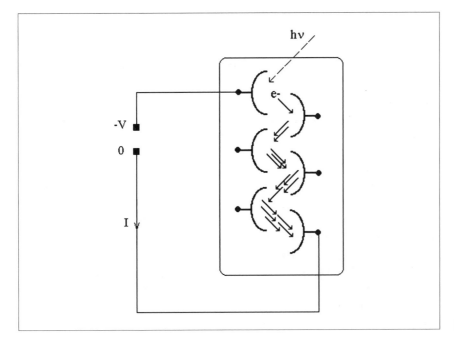

**Figure 2.9** Photomultiplier.

(or 'cascade') of photo-emissive electrodes under high vacuum, with a clear glass (or quartz) window for the light to enter. The electrodes are under successively higher voltage (up to around 1200 volts), so that any electrons emitted by one electrode are accelerated to hit the next. When a photon of sufficient energy strikes the first electrode in the series, a photoelectron is produced which is then accelerated towards the next electrode in the cascade. Collision of this electron with the electrode ejects further electrons, and so on down the cascade to produce a measurable large pulse of electrons at the final electrode (hence the term 'photomultiplier'). For relatively intense light, this may be measured as a continuous electric current. Alternatively, for low light detection levels, the photomultiplier may be operated in 'photon counting' mode whereby the discrete electronic pulses produced by single photons can be counted electronically. The quantum efficiency of such systems is very high, and detection of single photons is frequently only limited by the thermal 'noise' in the system (thermionic electron emission from the electrodes, *etc*.). This can be reduced by cooling the photomultiplier to low temperatures.

One disadvantage of photomultipliers is that they do not have spatial resolution, so the spectrum must be scanned to measure light intensity at each wavelength in turn. This can be overcome using a **diode array** or **charge-coupled devices** (CCDs) in which the intensity distribution across the entire wavelength field of the spectrometer may be captured simultaneously.

Photo-diodes are semiconductor devices whose electrical resistance is reduced when exposed to light. This can be detected by suitable electronics. Modern fabrication methods allows construction of arrays of such diodes (in one- or two-dimensions), each of which may be interrogated separately. This allows simultaneous measurement of light intensity over the array area.

Charge-coupled devices are efficient, silicon-based semiconductor photoelectric devices in which photon absorption creates photo-electrons or electron–hole pairs in the semiconductor, leading to the accumulation of charge at specific locations ('pixels') on the silicon wafer. Using electronic methods, this charge is transferred to sensor electrodes and measured one pixel at a time to build up an electronic image of the light intensity across the array. (These devices are now widely used as the imaging elements in digital cameras.)

Spectrophotometers based on array detectors differ slightly in configuration from those described above. In particular, the dispersive element or monochromator comes after the sample, so that the entire spectrum emerging from the sample may be imaged onto the detector array (Figure 2.10).

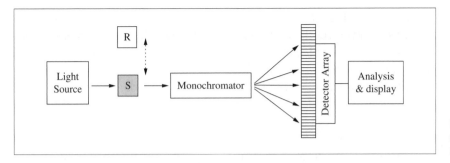

**Figure 2.10** Single beam spectrophotometer with array detector.

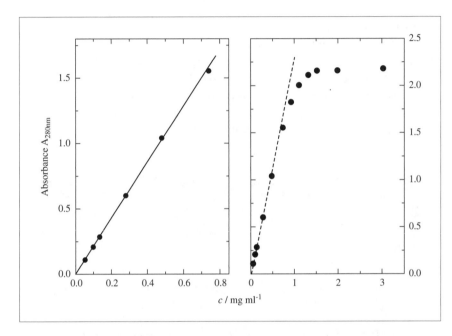

**Figure 2.11** Absorbance at 280 nm as a function of concentration for a typical protein solution, showing linear behaviour at low concentrations (left) but deviations from Beer–Lambert behaviour due to stray light effects at higher absorbances.

## 2.2.2  Experimental Limitations: The Stray Light Problem

As with any laboratory instrument, there are fundamental limitations in the accuracy of the information given by spectrophotometers. Even though the absorbance reading may appear quite precise (especially with digital and/or computer display), how do we know it is telling the truth?

Wavelength and absorbance calibration can be checked using standard filters. But, even with a well-calibrated instrument and properly prepared samples, things can go wrong. Look at the graph in Figure 2.11, which shows the actual measured absorbance at 280 nm for a series of standard protein solutions of different concentrations.

At low concentrations the graph of $A$ *vs. c* is quite linear, and obeys the Beer–Lambert law. However, at higher concentrations, when the absorbance approaches two or more, we begin to see values *lower* than

Remember the difference between **precision** and **accuracy**? A digital quartz watch may display the time precisely to fractions of a second, but will still tell the wrong time if not set properly. A correctly adjusted analogue watch (with hands!) may only display the time to the nearest minute, but it can still be the more accurate.

Deviations from the Beer–Lambert law (both positive and negative) can also occur because of concentration-dependent properties of the sample such as monomer–dimer equilibrium, which can alter spectral properties.

we should expect, eventually reaching a plateau above which the absorbance apparently remains constant regardless of concentration. This is an example of the 'stray light' effect in spectrophotometers and is a major instrumental limitation.

This stray light effect can be visualized as follows. When measuring the transmitted light ($I$), we assume that all the light reaching the detector is of the correct wavelength and has actually passed through the sample. But what if the sample compartment is not properly light-proof, or if the monochromator allows through some light of a different wavelength that is not absorbed by the sample? In such instances, the amount of light hitting the detector is greater than it should be. Consequently it appears to the measuring circuitry that the absorbance of the sample is less than it truly is. Most routine spectrophotometers are usually reliable up to $A = 1.5$, but rarely, except in specialist instruments, can one measure values exceeding $A \approx 2$ with any confidence. Samples must be chosen so that measurements lie within appropriate instrumental limits.

---

### Worked Problem 2.4

**Q:** The visual pigment chromophore, retinal (vitamin A aldehyde), has a molar extinction coefficient (in ethanol) of 43 000 at 375 nm.

(a) What should be the absorbance at 375 nm of a 45.0 $\mu$mol dm$^{-3}$ solution of retinal in a 1 cm cuvette?

(b) If due to stray light problems in the spectrophotometer, 0.5% of the light reaching the detector had not passed through the sample, what would be the measured absorbance? Assume that this spectrophotometer has a display reading to three decimal places.

(c) What would the results be for a 20 $\mu$mol dm$^{-3}$ solution in the same instrument?

**A:** (a) Using the Beer–Lambert law:

$$A_{375} = \varepsilon_{375}cl$$
$$= 43000 \times 45 \times 10^{-6} \times 1$$
$$= 1.935$$

(b) In the absence of stray light, the proportion of the light reaching the detector should be:

$$I = I_0 \times 10^{-A} = 0.0116 \, I_0$$

However, an additional $0.005 \, I_0$ (0.5%) is reaching the detector from other sources, so the actual light entering the detector is $I(\text{measured}) = (0.0116 + 0.005)I_0 = 0.0166 \, I_0$

Consequently, the measured absorbance would be:

$$A(\text{measured}) = -\log(I/I_0) = 1.780.$$

This corresponds to an almost 9% error (low reading).

(c) For a $20 \, \mu\text{mol dm}^{-3}$ retinal solution, the true $A_{375} = 43\,000 \times 20 \times 10^{-6} \times 1 = 0.86$ and $I = 0.138 \, I_0$

$$I(\text{measured}) = (0.138 + 0.005) \, I_0 = 0.143 \, I_0$$
$$A(\text{measured}) = 0.845$$

This corresponds to less than 2% error in the measured absorbance under these conditions. This emphasizes the need to be aware of specific instrument limitations.

## 2.2.3  Electronic Absorption: UV/Visible Spectra

UV and visible absorbance spectra arise from electronic transitions in atomic or molecular orbitals (Figure 2.12).

## 2.2.4  Spectral Characteristics of Biomolecular Chromophores

Most biological macromolecules are colourless to the human eye and only reveal their spectral characteristics when viewed in the UV range. Absorption spectra are characterized by their shape, the peak wavelength ($\lambda_{\max}$), and the peak height or molar extinction coefficient ($\varepsilon$). Spectra of amino acids, nucleic acids, proteins, DNA and other chromophores are described in the following section.

### 2.2.4.1  Amino Acids, Peptides and Proteins

A typical UV spectrum of a simple globular protein in solution is shown in Figure 2.13. This particular protein contains no groups other

Absorbance spectra of chromophores in solution are usually much broader and contain less fine structure than molecules in the gas phase. This is because of broadening effects arising from the collisions and other interactions with nearby groups and solvent molecules. First, the slightly different and fluctuating molecular environments experienced by chromophores will give rise to different energy level spacings and broader vibrational bands (**heterogeneous broadening**). Secondly, collisions with surrounding molecules will reduce the lifetime of the excited state and, following the Heisenberg uncertainty principle that $\Delta E.\Delta t \geq \hbar$ also lead to broader spectral bands (**homogeneous** or **lifetime broadening**).

The word 'chromophore' derives from the ancient Greek *chroma* (colour) *phoros* (bearing). It is the term used to signify any group or part of a molecule that is coloured or has UV/visible absorbing properties.

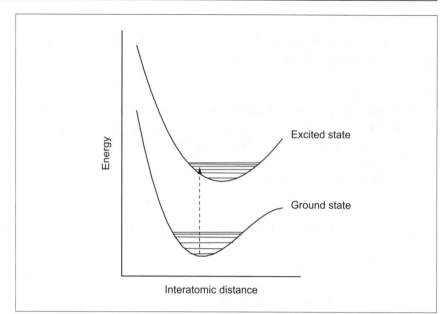

**Figure 2.12**   Absorption of a photon of appropriate energy can promote an electron from a lower energy ground state into a range of closely spaced excited states, from where it may decay back to the ground state, usually dissipating the energy as heat.

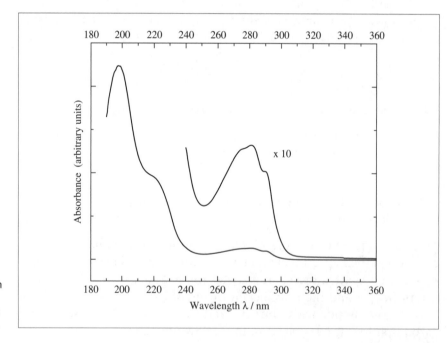

**Figure 2.13**   Typical UV spectrum for a globular protein in aqueous solution. The near-UV region (>240 nm) is also shown for clarity on an expanded scale.

than the constituent amino acids, so all the spectral characteristics arise from the amino acid groups themselves. The absorbance at very short wavelengths (<240 nm) is due mainly to $\pi$–$\pi^*$ transitions of the peptide amide and related groups, and is not normally of much direct interest (though see circular dichroism in Section 2.3). The more useful near-UV region (Figures 2.14 and 2.15, Table 2.1) shows the characteristic absorbance spectrum arising from a combination of aromatic side chains (tryptophan, tyrosine, phenylalanine) and, to a lesser extent, cysteine residues.

One important application of UV/vis spectroscopy is in measuring concentrations—either using intrinsic chromophores or by observing the coloured product of a suitable reaction.

If the amino acid composition of the protein is known, then we can use the molar extinction coefficients of the chromophoric side chains to estimate the extinction coefficient of the entire protein.[1] This is

See Chapter 1 for the structures of the aromatic side chains.

Accurate measurement of protein concentrations in particular is surprisingly difficult and a number of methods have been devised. However, except in special circumstances, it is rarely possible to estimate protein concentrations in solution to better than ±5% accuracy.

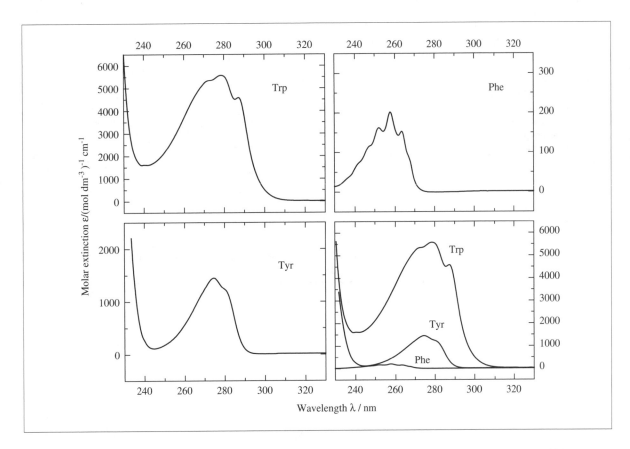

**Figure 2.14** Near-UV absorbance spectra for aromatic amino acid side chains. When superimposed on the same scale (bottom right panel), the dominant contribution from tryptophan and tyrosine residues becomes more obvious.

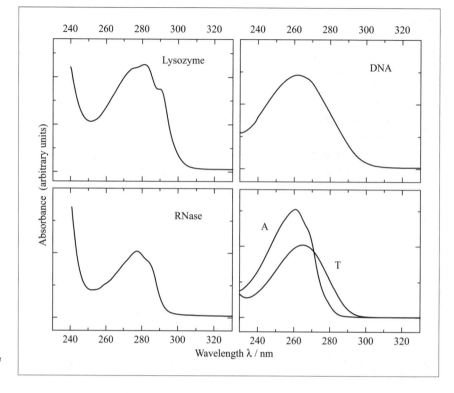

**Figure 2.15** Typical near-UV absorbance spectra for proteins (lysozyme, ribonuclease) and nucleic acids (DNA, adenine, thymine). Notice how the lysozyme spectrum (six Trp, three Tyr, three Phe) is dominated by the tryptophan absorbance, whereas ribonuclease (RNase: zero Trp, six Tyr, three Phe), which (unusually) contains no tryptophan residues, is more characteristic of the tyrosine side chains.

**Table 2.1** Approximate $\lambda_{max}$ and $\varepsilon$ for biological chromophores in the near-UV/visible region under physiological conditions

|  | $\lambda_{max}$/nm | $\varepsilon/(mol\,dm^{-3})^{-1}\,cm^{-1}$ |
|---|---|---|
| **Nucleotides:** | | |
| A | 259 | 15 400 |
| C | 271 | 9200 |
| G | 252 | 13 700 |
| U | 262 | 10 000 |
| T | 267 | 10 200 |
| **Amino acids:** | | |
| Trp | 280 | 5690 |
| Tyr | 276 | 1400 |
| Phe | 257 | 160 |
| Cys | <250 | 60 (at 280 nm) |
| **Others:** | | |
| retinal (in rhodopsin) | 500 | 42 000 |
| haem (in haemoglobin): oxy- | 414 | 131 000 |
| deoxy- | 432 | 138 000 |
| NAD | 260 | 18 000 |
| NADH | 340 | 6200 |
| chlorophyll | 418 | 111 000 |

normally done at 280 nm, where the accepted $\varepsilon_{280}$ values are: 5690 (Trp), 1280 (Tyr) and 60 (Cys—half cysteines only). Phe does not absorb significantly at 280 nm.

---

**Worked Problem 2.5**

**Q:** A globular protein of relative molecular mass 23 500 contains six Trp, four Tyr and three Phe residues.

(a) What is the $\varepsilon_{280}$ for this protein?

(b) What would be the absorbance at 280 nm for a 0.5 mg cm$^{-3}$ solution in a 1 cm cuvette?

**A:** (a) $\varepsilon_{280} = 6 \times 5690(\text{Trp}) + 4 \times 1280(\text{Tyr}) + 0(\text{Phe}) = 39\,260\,\text{mol}^{-1}\,\text{dm}^3\,\text{cm}^{-1}$

(b) Molar concentration: $0.5\,\text{mg cm}^{-3} \equiv 0.5\,\text{g dm}^{-3} \equiv 0.5/23\,500 = 2.13 \times 10^{-5}\,\text{mol dm}^{-3}$

$$A_{380} = \varepsilon_{380}cl = 39\,260 \times 2.13 \times 10^{-5} \times 1 = 0.836$$

---

Various colorimetric methods are also available based on non-specific dye binding to polypeptide chains, one of the more common being the 'Bradford' assay.[2] One drawback with such methods is that the actual colour intensity (absorbance) developed is not absolute, but depends on the specific protein. Calibration can therefore be a problem if accurate concentrations are required.

Strictly speaking, each of these methods measures only the polypeptide concentration, taking no account of whether the protein is correctly folded or active.

### 2.2.4.2 Nucleic Acids

Nucleic acids (DNA, RNA) have strong UV absorbances around 260–270 nm arising from $\pi$–$\pi^*$ transitions in the aromatic purine (A, G) and pyrimidine (U, C, T) rings of the nucleotide bases. The sugar phosphate backbone groups absorb only in the far UV.

DNA and RNA spectra are quite sensitive to conformation, since the regular stacking of the nucleotide chromophores in helical structures leads to the phenomenon of '**hypochromicity**'.

Because of base-stacking interactions that perturb the spectroscopic properties of the closely packed aromatic rings, helical polynucleotides

The Bradford assay (named after its inventor[2]) uses Coomassie brilliant blue dye, a histological stain which binds to protein and, as the name suggests, turns a bright blue colour. At low concentrations, the absorbance is proportional to the protein concentration in solution.

Reminder: the structures of adenosine (A), guanine (G), uracil (U), cytosine (C) and thymine (T) are given in Chapter 1.

Greek: *hypo-* = lower, *hyper-* = greater.

The melting and refolding of double-helical DNA is the basis of the polymerase chain reaction (PCR) method now widely used in molecular biology for amplifying specific DNA sequences, and for which Kary Mullis won the 1993 Nobel Prize in chemistry.

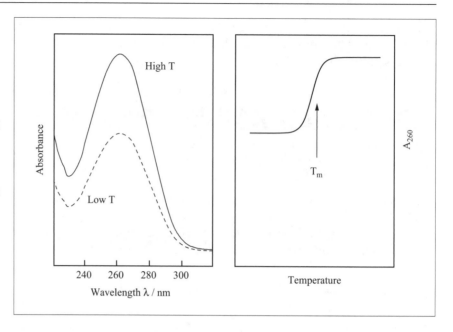

**Figure 2.16** Hypochromism and the 'melting' of DNA. The left-hand panel shows the UV spectra of a solution of DNA at a low temperature (25 °C, say) compared to the same solution at a higher temperature (typically >60 °C). The difference is due to the unstacking of the base pairs as the DNA double helix unfolds (or 'melts') at higher temperatures. This can be used to follow the unfolding process with temperature (right-hand panel).

have a lower absorbance in the 260 nm region than would be expected from the free nucleotides alone. In particular, double-stranded helical DNA has a lower $A_{260}$ than the denatured single strands. The increase in absorbance can amount to as much as 40%, and this is very useful for measuring the thermal unfolding or 'melting' of polynucleotides in solution (Figure 2.16).

### 2.2.4.3  Prosthetic Groups and Other Chromophores

In addition to the intrinsic chromophores of proteins and nucleic acids, there are a number of other co-factors and prosthetic groups that have characteristic absorbance properties in the UV and visible region. For example, the familiar colour of green plants is due primarily to the absorbance of red light by chlorophylls and other photosynthetic pigments, mainly bound in specific protein structures. The mammalian haem proteins (haemoglobin, myoglobin) have a characteristic red colour arising from absorption at the blue end of the spectrum by the iron–porphyrin prosthetic group (haem) also responsible for oxygen binding.

The spectral properties of prosthetic groups are frequently modified by covalent and non-covalent interactions with their host protein. For example, the major chromophore of the visual pigment protein rhodopsin is retinal (vitamin A aldehyde). This is a conjugated polyene molecule which is incorporated into the protein, usually (in the dark) as the 11-*cis* isomer and which, on exposure to light, is photoisomerized to

The familiar rusty-brown colour of old bloodstains is due to oxidation of iron(II) to iron(III) in the haem group of the protein haemoglobin. Everyday experience tells us that iron, in combination with oxygen and water, reacts easily to form iron(III) oxides or 'rust'. This is very hard to reverse and would be literally fatal if it were to occur to any great extent in our haemoglobin, myoglobin or other haem-containing oxygen transport proteins. One of the remarkable properties of such proteins is the way in which the macromolecular structure is so arranged as to allow easy access for binding and release of $O_2$ to the iron(II) while at the same time excluding the water that might catalyse oxidation. When the protein is denatured (as in old blood), this protection is lost.

the all-*trans* form as the first step in the visual process. But retinal in its free state absorbs only in the near UV (around 370 nm) and, as such, would be ineffective for detecting light in the visible region. However, when bound to the protein it forms a covalent protonated imine (Schiff base) linkage to a specific lysine side chain which, together with non-covalent interactions with other amino acids in the binding site, shifts the absorbance $\lambda_{max}$ into the 400–600 nm visible region.

The oxidized and reduced forms of important electron transfer co-factors such as nicotinamide adenine dinucleotide (NAD) differ in their UV absorbing properties (see Table 2.1). This is very useful in following the redox state of these prosthetic groups during metabolic pathways, and the UV absorbance changes are often used to advantage in developing methods for measuring the kinetics of enzyme-catalysed reactions involving such groups.

### 2.2.4.4   Lipids and Carbohydrates

Most lipids and carbohydrate molecules lack the extensive electron delocalization or aromatic groups necessary for the relatively low energy electronic transitions required for near UV or visible absorbance. Consequently they usually absorb only at very short wavelengths (<200 nm) except when chemically degraded (as, for example, in the 'caramelization' of sugars). Moreover, the insolubility of lipids in water and their tendency to form bilayers and other aggregates gives rise to light scattering, particularly at shorter wavelengths, which makes it difficult to measure their true absorbance properties.

### 2.2.5   Factors that can Affect UV/Visible Spectra

The factors that determine the wavelength and intensity of UV/visible absorption are complex and we have already seen above how base stacking interactions can modify the absorbance of nucleic acids (the **hypochromic effect**).

The $\lambda_{max}$ depends primarily on the available electronic molecular energy levels for a particular chromophore, whereas $\varepsilon$ is determined by its absorption cross-section ($\sigma$). Although both of these are intrinsic quantum mechanical properties of the molecular orbitals, both can be affected by the surrounding molecular environment and there are some simple physical considerations that can be helpful.

Excitation of valence electrons to higher energy levels inevitably involves separation of negative charge (the electron) from the positively charged atomic nuclei. Consequently there is a change in (electric) dipole moment in going from ground to excited state, and excited states normally have a higher dipole moment (known as the **transition dipole**).

The **absorption cross-section** ($\sigma$) is related to the probability that a photon will be absorbed when encountering the chromophore. It can be pictured as the cross-sectional area of a hypothetical opaque object equivalent to the absorbing molecule (see Section 2.1.2).

Simple electrostatics dictates that the work needed to separate these charges will depend on the polarity or dielectric constant (relative permeability) of the surroundings. For example, if a chromophore is surrounded by a very polar solvent such as water, the energy required to form the excited state dipole will be less than for the same process in a non-polar environment. As a result, the energy level differences will be slightly smaller for the chromophore in water and the absorption spectrum might be expected to be shifted to lower energies/ longer wavelengths. Similar considerations apply to the intensity of absorbance.

These effects can be very useful in following conformational changes or binding processes in biological macromolecules. For example, when a globular protein unfolds, the aromatic side chains (Trp and Tyr), which are usually partly buried within the relatively non-polar interior of the protein, become exposed to the more polar aqueous environment. This brings about small changes in the UV absorbance spectrum in the 270–290 nm region which can be used to determine the extent of unfolding under specific conditions. Changes in UV spectra may also come about when something binds in the active site of a protein. For example, the binding site cleft of the enzyme lysozyme contains a number of tryptophan side chains that are partially exposed to water. When a trisaccharide inhibitor is added to the solution, it binds in this active site and reduces the polarity of the Trp environment, bringing about a small shift in the overall spectrum.

### 2.2.5.1  Difference Spectra

Because these changes in UV/vis spectra are relatively small, they can often be difficult to see directly from the spectra themselves. In such cases difference spectra can be useful. This involves subtraction of the original spectrum from the spectrum of the perturbed sample. Often this is best done directly in the spectrophotometer by loading a portion of the original sample in the reference cuvette (rather than solvent) and performing the perturbations on the same material in the sample cuvette (Figure 2.17).

The related techniques of **optical rotation** and **optical rotatory dispersion** (ORD) (*i.e.* optical rotation as a function of wavelength) are less frequently used nowadays for studying biomolecules.

## 2.3    Circular Dichroism

Chiral molecules may respond differently to left- or right-circularly polarized light. In particular there may be slight differences in absorbance of UV/visible light; these provide the basis for circular dichroism (CD) spectroscopy.

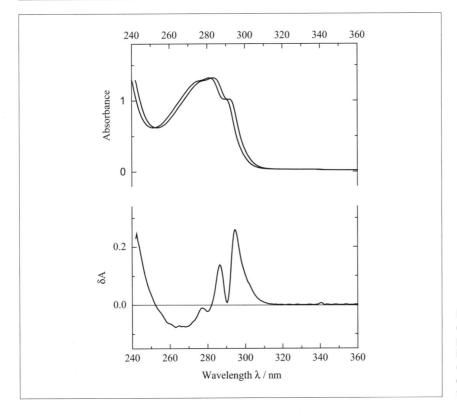

**Figure 2.17** Example of difference spectrum. The upper panel shows two (almost identical) protein spectra. The difference spectrum ($\delta A$) obtained by subtraction highlights the differences more clearly.

Circular dichroism is measured as the difference in absorbance between left and right circularly polarized light:

$$\Delta A = A_L - A_R$$

where $A_L$ and $A_R$ refer to the absorbance of the sample at a particular wavelength for left- or right-circularly polarized light, respectively (Figure 2.18). This can be expressed in terms of differences in molar extinction coefficients, $\Delta\varepsilon_{molar} = \varepsilon_L - \varepsilon_R$, which for historical reasons is often converted to **ellipticity**, $\theta$. (See Box 2.2 for more details.)

CD spectra are measured in an instrument ('spectropolarimeter'), which is, in principle, similar to a single-beam spectrophotometer, but using circularly polarized light (Figure 2.19). But because CD effects are usually quite small and observations need to extend into the far-UV, much more intense light sources are required and the instruments are rather more specialized.

Normal linear- or un-polarized light is made up of a superposition of circularly polarized photons of different handedness. Left- or right-polarization can be selected using an electro-optic modulator (or photoelastic modulator) in which a piezoelectric crystal of quartz, or other optically active material, is subjected to an alternating electric

Light is said to be circularly polarized when the oscillating electric field vector rotates (either to the left or to the right) about the propagation axis of the electromagnetic wave, so that the tip of the vector follows a helical path along the beam. (The magnetic vector will do the same, but at right angles to E.) Most light is unpolarized—in reality a random mixture of all possible polarizations. Linear- or plane-polarized light is made up of a superposition of left- and right-circularly polarized light in equal proportions. Light made up of unequal proportions of left- and right-circularly polarized components is said to be **elliptically polarized**. This is why CD data are often quoted in terms of the ellipticity, $\theta$.

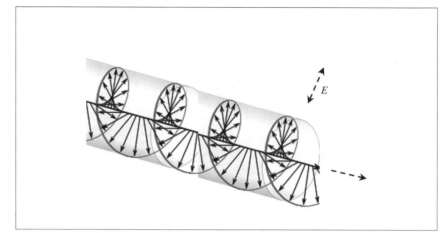

**Figure 2.18** Helical path traced out by the electric vector in circularly polarized light. The clockwise rotation shown here corresponds to right circular polarization. Left circularly polarized light would be anti-clockwise in this representation.

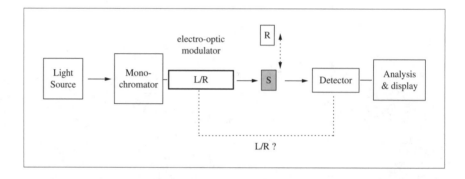

**Figure 2.19** Sketch showing the different components of a typical spectropolarimeter.

Another way of looking at this is to remember that, in quantum theory, light is made up of photons, particles of spin-1, in which the direction of the spin axis is either parallel or anti-parallel (pointing forwards or backwards) to the direction of travel. These two spin states correspond to left or right circular polarization.

field which distorts the crystal in such a way as to transmit left or right circular polarized light alternately. This passes through the sample and into the detector and analysis circuitry, which are electronically synchronized with the modulator so that light transmitted in left- or right-polarization can be measured separately. The cuvettes for liquid samples are of similar construction to those used for normal UV spectroscopy, though frequently of much shorter path lengths for use in the strongly absorbing far-UV range.

## Box 2.2 Circular Dichroism Units

The units used for reporting circular dichroism data can sometimes be confusing. The quantity measured ($\Delta A = A_L - A_R$) is in dimensionless absorbance units, and follows all the rules and experimental restrictions that apply to related absorbance measurements (Section 2.1). In particular, it is often convenient to use the *molar* differential extinction coefficient.

$$\Delta\varepsilon_{molar} = \varepsilon_L - \varepsilon_R$$

This would be the differential absorbance seen for a hypothetical 1 M ($1\,mol\,dm^{-3}$) solution with path length of 1 cm. The units of $\Delta\varepsilon_{molar}$ are $mol^{-1}\,dm^3\,cm^{-1}$, where the concentration ($1\,mol\,dm^{-3}$) is expressed per mole of (macro)molecule in solution.

For proteins and polypeptides in the far-UV region (180–240 nm), the majority of the absorbance arises from the peptide groups. Consequently, for ease of comparison, data are often presented per mole of peptide unit, or per mean residue weight (MRW). A protein made up of a single chain of $N$ amino acids will contain $N-1$ peptide units. The mean residue weight will be:

$$MRW = RMM/(N-1)$$

where RMM is the relative molar mass (in Da) of the protein. So, in terms of peptide units or mean residue weight:

$$\Delta\varepsilon_{MRW} = \Delta\varepsilon_{molar}/(N-1).$$

For most proteins, MRW is about 110 Da. This value can be used to estimate MRW concentrations (or $N$) in situations where the size of the protein is not known and the concentration is estimated by weight (*e.g.* in $mg\,mL^{-1}$).

Similar considerations also apply to nucleic acids and other biopolymers, where the CD effect is often conveniently expressed per mole of repeating unit.

For historical reasons related to the phenomenon of optical rotary dispersion (ORD), CD data are often reported in terms of the ellipticity, $\theta$. Plane-polarized light passing through a CD-active sample will become elliptically polarized because of the differential absorbance of the left- and right-circular components of the original beam. The ellipticity is the angle defined by $\theta = \tan^{-1}(b/a)$, where $a$ and $b$ are the lengths of major and minor axes, respectively, of the polarization ellipse.

Mean residue ellipticities and differential absorbances are related by a simple numerical factor:

$$\theta_{MRW} = 3298 \times \Delta\varepsilon_{MRW}, \text{ in units of degree } cm^2\,dmol^{-1}.$$

Typical CD spectra for proteins and nucleic acids are shown in Figures 2.20 and 2.21. It is important to remember that only chiral samples can give rise to circular dichroism or other optical activity phenomena. Because circular dichroism is made up of a difference

For most biological molecules in solution, the CD effect is quite small, with $\Delta A$ typically of the order of $3 \times 10^{-4}$, corresponding to ellipticities of around 10 millidegrees (mdeg).

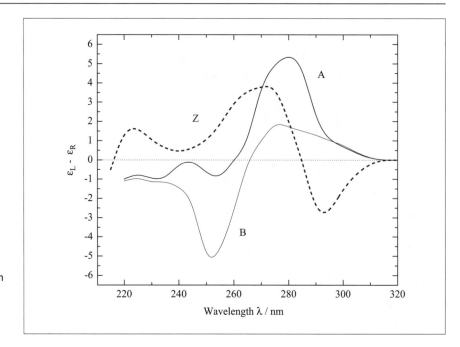

**Figure 2.20** Circular dichroism spectra showing mean residue differential absorbances for the A, B and Z forms of double-helical DNA.

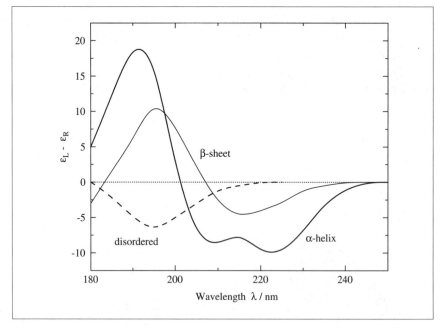

**Figure 2.21** Characteristic shapes for the CD spectra of some different types of secondary structure conformations found in polypeptides and proteins. The magnitudes of the mean residue differential absorbances ($\varepsilon_L - \varepsilon_R$, $mol^{-1}\ dm^3\ cm^{-1}$) are typical for proteins.

between absorbances, a CD spectrum may be positive or negative depending on wavelength, and the shape of the spectrum is generally characteristic of the conformation of the molecules involved.

The purine and pyrimidine bases of nucleic acids are not intrinsic-ally chiral, but DNA and RNA do nonetheless give significant CD spectra in the near and far-UV regions. This is because the double helix and other chiral conformations place the bases in a chiral environment that is reflected in their optical response to polarized light. This is illustrated in Figure 2.20, which shows the characteristically different CD spectra observed with the A, B and Z forms of DNA. The B-form is the right-handed double-helical conformation that is most commonly present under physiological conditions. If the water content of the sample is reduced (typically by adding alcohols or by dehydrating DNA gels), then the conformation switches to the A-form, with the base-pairs tilted with respect to the helix axis. The Z-form of DNA is a much rarer left-handed helical conformation, usually seen at high salt concentrations and first observed experimentally from its markedly different CD spectrum.[3]

Proteins are made up of amino acids which are (with the exception of glycine) intrinsically chiral. In addition, secondary structure elements such as α-helix and β-sheet impose additional chirality on the polypeptide chain. This gives rise to CD effects in the far UV (190–240 nm) arising mainly from the peptide backbone groups which are characteristic of the overall secondary structure of the protein (Figure 2.21). In the near-UV range (240–300 nm), induced (conformational) chirality in chromophoric residues such as tryptophan and tyrosine can give rise to CD effects in this region—although the side chain groups are not inherently optically active. Prosthetic groups or other chromophoric ligands bound to the protein may also show CD effects in the UV/vis region, either because they are intrinsically optically active, or because of chirality induced upon the molecules when closely bound to the optically active protein. All these effects will be sensitive to the secondary or tertiary structure of the protein, and CD has proved to be a very useful analytical technique for estimating conformation and for following conformational change.[4]

Methods are available to estimate the secondary structure content of proteins from their characteristic CD profiles. These algorithms use a library of known proteins with well-defined secondary structures to estimate the proportions of α-helix, β-sheet, β-turn and other conformational elements.

## 2.4   Fluorescence

After photon absorption to (singlet) electronically excited states, most atoms and molecules decay non-radiatively back to the ground state, transferring the excess energy to the surroundings generally as heat. Some molecules, however, display fluorescence in which a (small) proportion of excited molecules return to the ground state with the emission of a photon (Figure 2.22).

It was the A-form of DNA that was first observed experimentally in the original X-ray fibre diffraction experiments conducted by Rosalind Franklin working at King's College London, at the same time that Watson and Crick (in Cambridge) were devising their double-helical model of DNA. She was experimenting with drawn fibres of DNA exposed to the air, which naturally tended to dry out. Later, as techniques developed, the fibres were enclosed in glass capillaries in which the moisture content could be better controlled.

The term 'fluorescence' was coined in 1852 by G. G. Stokes, after the mineral fluorspar ($CaF_2$), which emits visible light when illuminated with UV.

The **Frank–Condon principle:** since atomic nuclei are much heavier than electrons, electronic transitions usually occur vertically (Figure 2.22) before the nuclei have time to move. Changes in bond length take place more slowly.

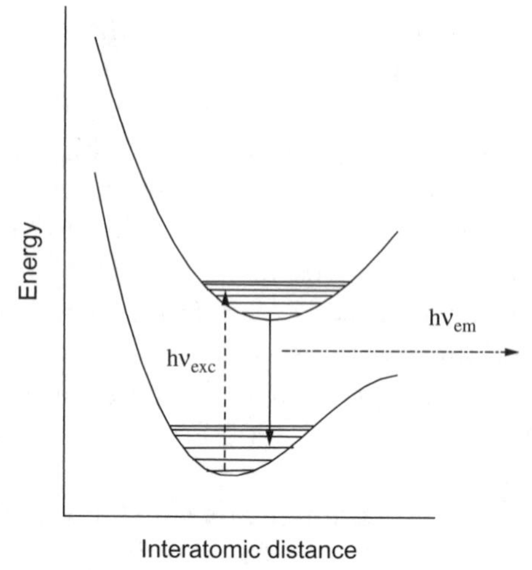

**Figure 2.22** Energy level diagram illustrating electronic excitation followed by fluorescence emission. After initial (vertical) excitation from the ground state, the system rapidly relaxes to the energy minimum of the excited state.

In **fluorescence microscopy**, the spectrofluorimeter becomes part of the microscope optics, frequently using optical filters instead of the more cumbersome monochromators. Detection systems may use video cameras or other area detectors, possibly with image intensifiers to enhance the intrinsically weak signal from microscopic samples.

Fluorescence radiation is emitted in all directions. It is characterized by the **quantum yield** ($\phi$) or intensity of the fluorescence and by its maximum wavelength ($\lambda_{max}$). For energy conservation reasons, the emitted photon must always be of equal or lower energy (longer wavelength) than the exciting light.

Fluorescence is normally measured using a **spectrofluorimeter**. This is made up of a light source and excitation monochromator to provide the light beam of selected wavelength to focus on the sample, followed by an emission monochromator and detector to monitor the light emitted from the sample, usually at right angles to the exciting beam so as to minimize the effects of scattered light (Figure 2.23).

The light source is usually a high intensity xenon arc lamp, though for more specialized applications a laser or tunable laser might be used instead (in which case the excitation monochromator is not needed). The detector might be a photomultiplier or array detector, as described in Section 2.2.

Samples for fluorescence conventional experiments (usually solutions) are placed in a rectangular cuvette with four optically clear faces. Light from the excitation monochromator, at a wavelength ($\lambda_{exc}$) within the absorbance band, is focused on the centre of the sample. Any light coming at right angles (from fluorescence or any other source) is focused onto the entrance slit of the second (emission) monochromator so that the spectrum of the emitted light may be measured ($\lambda_{em}$).

An **emission spectrum** is obtained by exciting the sample at fixed wavelength ($\lambda_{exc}$) and observing the spectrum of the emitted light. Alternatively, for an **excitation spectrum**, the range of excitation

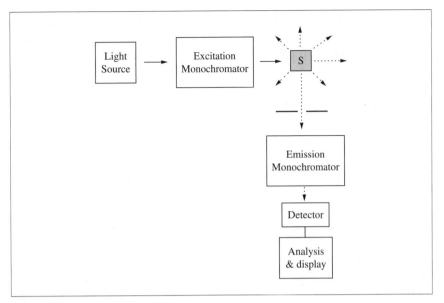

**Figure 2.23**  Sketch of a typical spectrofluorimeter.

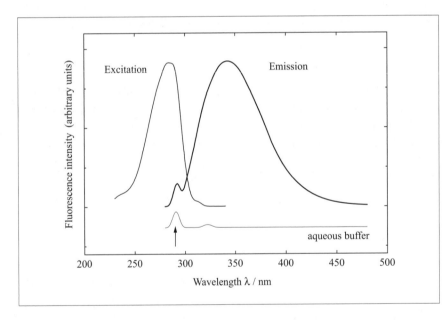

**Figure 2.24**  Typical fluorescence excitation and emission spectra for a globular protein in aqueous buffer at room temperature. The excitation wavelength, $\lambda_{exc}$, is 290 nm (arrow). The emission spectrum baseline measured with buffer in the absence of protein is shown offset for clarity.

wavelengths is scanned while observing the light intensity emitted at fixed $\lambda_{em}$. Since the light intensity from the source will vary with wavelength, for accurate quantitative work it is necessary to correct the observed spectra for these variations in source intensity.

Typical fluorescence excitation and emission spectra are shown in Figure 2.24.

Points to note in Figure 2.24 are that the excitation spectrum is very similar to the absorbance spectrum, as it should be since fluorescence can only occur if the sample actually absorbs photons at the excitation wavelength. The emission spectrum, on the other hand, comes at longer wavelengths, and is roughly a mirror image of the excitation spectrum. This is characteristic of all fluorescence spectra. The large emission intensity at longer wavelengths, peaking here at about 342 nm, is characteristic of the strong fluorescence from tryptophan residues in the protein.

The baseline signal from the buffer alone highlights the presence of two additional features which are not from fluorescence, but which must be accounted for. First, there is a (relatively) small peak at the same wavelength as the exciting light. This is simply due to scattered light from the sample and corresponds to photons that have been elastically scattered (*i.e.* without loss of energy) from the sample molecules or from dust or other scattering contaminants. A second small feature, peaking in Figure 2.24 at about 320 nm, arises from the Raman inelastic scattering of photons from the solvent water. The energy loss in the Raman scattered photons corresponds to the vibrational energy excited in the water molecules as they interact with the incoming photons. (See Section 2.5 for more on Raman spectroscopy.)

### 2.4.1  Inner Filter Effects

Spurious fluorescence spectra may sometimes be seen with samples that are too concentrated. This is due to the inner filter effect, in which a significant proportion of the excitation beam is absorbed by the sample before it reaches the centre of the cuvette (where the collection optics are focused), or the emitted light is similarly re-absorbed before it gets out of the cuvette. A useful rule of thumb is that the absorbance ($A$) of the sample at the excitation/emission wavelengths should be less than 0.2.

### 2.4.2  Fluorescence Shifts

One of the more useful aspects of fluorescence as a biophysical tool is the way in which both fluorescence intensity and wavelengths are sensitive to chemical and environmental effects.

The wavelength of maximum emission ($\lambda_{max}$) is determined by the energy difference between the ground and excited states of the fluorescent molecule, and this can be affected by the polarity of the surrounding solvent in exactly the same way as described for absorbance (see Section 2.2.5). An example of this is given in Figure 2.25, which compares the fluorescence emission spectra of tryptophan in different solvents. In general, because of the excited state transition dipole effect, a more polar environment will reduce the energy separation and

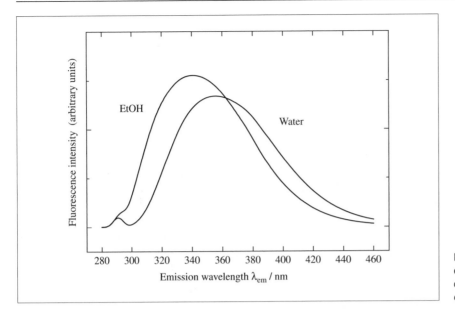

**Figure 2.25** Fluorescence emission spectra ($\lambda_{exc} = 290$ nm) of tryptophan in water and in ethanol solutions.

will give a more red-shifted fluorescence emission spectrum (larger $\lambda_{max}$). On the other hand, a non-polar environment will tend to give a blue-shifted spectrum (shorter $\lambda_{max}$). Such qualitative effects can be very useful in interpreting conformational and interaction properties of macromolecules.

### 2.4.3 Fluorescence Intensity: Quantum Yield and Quenching

The fluorescence quantum yield, $\phi$, is equal to the number of photons emitted divided by the number of photons absorbed. It can be related to the absorbance properties of the molecule and the **fluorescence lifetime** ($\tau$) as follows:

$$\phi = \tau / \tau_0$$

where $\tau_0$ is the fluorescence lifetime of the excited state in the absence of non-radiative processes. The overall fluorescence lifetime ($\tau$) is the total lifetime of the excited state, taking into account all possible decay modes and is typically up to 10 ns for most fluorophores.

The longer the molecule exists in the excited state, the more chance it has of fluorescing. Anything that reduces this lifetime by providing an alternative, non-radiative path back to the ground state will reduce $\phi$ and will lower the fluorescence intensity. This is known as **fluorescence quenching**.

The fluorescence quantum yield can be expressed in terms of the relative rates of radiative ($k_r$) and non-radiative ($k_{nr}$) decay of the excited state:

$$\phi = k_r/(k_r + k_{nr}) = 1/(1 + k_{nr}/k_r)$$

The combined rate of decay of the excited state is related to its lifetime:

$$k_r + k_{nr} \approx 1/\tau$$

In addition, there are a number of factors that may contribute to the rate of non-radiative decay ($k_{nr}$). These can be written as follows:

$$k_{nr} = k_{intrinsic} + k_{environmental} + k_{dynamic\ quench} + k_{static\ quench}$$

Think of the excited state electrons as objects on a shelf. The electrons will have a natural tendency to fall off the shelf at a rate ($k_{intrinsic}$) that will depend on the specific molecular structure. If, in addition, the shelf is being rattled by the continual Brownian motion bombardment of surrounding molecules and groups, then the electrons may be displaced by environmental interactions. The rate at which this occurs ($k_{environmental}$) will depend on the frequency of molecular collisions (temperature), and on the size and polarity of the colliding species. For example, more polar molecules in the surrounding solvent will tend to interact more readily with the excited state electrons due to electrostatic forces. So we might anticipate that fluorescence intensities will be reduced by increase in temperature or transfer to a more polar solvent.

There can be more specific quenching effects also. Some molecules or groups such as oxygen ($O_2$), iodide ($I^-$), acrylamide (propenamide) and succinimide (pyrrolidine-2,5-dione) will accelerate non-radiative decay if they come in contact with the excited fluorophore. Such quenching may be **dynamic**, involving transient collision between the molecules, or **static** if the quencher molecule (Q) forms a longer lived complex with the fluorescent group.

In the case of **dynamic quenching**, the rate depends on the frequency of collisions and will vary with quencher concentration:

$$k_{dynamic\ quench} = k_{dyn}[Q]$$

where $k_{dyn}$ is the pseudo-first order rate constant for diffusional collisions of the quencher, Q, with the excited group. Consequently, in the absence of specific quenching, we may write the fluorescence quantum yield (or intensity) as:

$$\phi_0 = k_r/(k_r + k_{intrinsic} + k_{environmental})$$

and in the presence of dynamic quenching:

$$\phi = k_r/(k_r + k_{intrinsic} + k_{environmental} + k_{dyn}[Q])$$

The ratio of fluorescence intensities in the absence or presence of a given quencher concentration can be written as:

$$\phi_0/\phi = 1 + K_{SV}[Q]$$

This is the simplest form of the Stern–Volmer equation, where the Stern–Volmer coefficient ($K_{SV}$) is given by:

$$K_{SV} = k_{dyn}/(k_r + k_{intrinsic} + k_{environmental})$$

**Static quenching** involves the formation of an equilibrium complex between the quencher (Q) and the fluorescent group (F):

$$F + Q \rightleftharpoons FQ; \qquad K = [FQ]/[F][Q]$$

where K is the association constant for the complex. Assuming that only the uncomplexed molecules can fluoresce, the relative fluorescence intensities in the absence or presence of static quencher will be given by:

$$\phi_0/\phi = ([F] + [FQ])/[F] = 1 + [FQ]/[F] = 1 + K[Q]$$

These equations relating dynamic or static quenching to quencher concentration are of the same form, and a plot of $\phi_0/\phi$ vs. [Q] will be linear in each case (Stern–Volmer plot, see Figure 2.26). Consequently it can be difficult to discriminate between these two quenching mechanisms. Dynamic quenching can be sensitive to viscosity since it depends on the rate of diffusional collisions, and comparison of Stern–Volmer plots from experiments in solvent mixtures with different viscosities can sometimes be helpful. In practice, however,

It was the observation of dynamic quenching by molecular oxygen of tryptophan residues buried within globular proteins that gave some of the first experimental indications that protein structures are not rigid, but undergo significant dynamic fluctuations.[5]

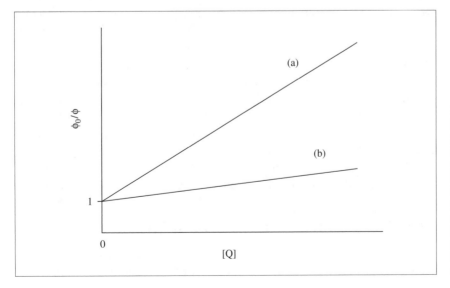

**Figure 2.26** Stern–Volmer plots for quenching of fluorescence in: (a) tryptophan in solution; (b) tryptophan residues in a globular protein, by quencher molecule, Q. The smaller slope ($K_{SV}$) for the protein indicates that the Trp residues are less dynamically accessible than the same groups in solution, consistent with burial of the Trp side chains within the folded conformation of the macromolecule.

Stern–Volmer plots are frequently non-linear, with the curvature indicating the presence of a number of different fluorophores in different environments and that a number of quenching mechanisms may be involved. Nonetheless, qualitative comparison can be useful in identifying the locations of fluorescent groups such as tryptophan residues in protein structures.

### 2.4.4  Intrinsic Fluorescence

Natural fluorescence and phosphorescence is common in many aquatic organisms. The **green fluorescent protein** (GFP) from a species of jellyfish, *Aequorea victoria*, is now widely used in molecular biology.[6] Its intrinsic fluorescence arises from a specific covalent rearrangement of a Ser-Tyr-Gly sequence within its structure.

Intrinsic fluorescence is relatively rare in biological molecules. Most of the naturally occurring nucleic acids, carbohydrates and lipids show little or no useful fluorescence in the normal UV/visible region. In proteins, fluorescence can only be seen from tryptophan residues and, to a lesser extent, tyrosine side chains. This intrinsic protein fluorescence can be used in a number of practical applications. Typical Trp and protein emission spectra are shown in Figures 2.24 and 2.25. Some proteins contain intrinsically fluorescent prosthetic groups such as reduced pyridine nucleotides and flavoproteins, and the chlorophylls from green plants show a red fluorescence emission.

### 2.4.5  Fluorescence Probes

To **intercalate** means to interpose or insert between, like placing a book between others on a library shelf. Intercalation of planar, aromatic molecules between neighbouring base pairs in DNA can disrupt the genetic machinery and this is why many such compounds are cancer suspect agents.

Because of the rarity of intrinsic fluorescent groups in biological molecules, it is frequently necessary to introduce unnatural fluorophores or fluorescent probes. A large number of such probes have been developed to cover a variety of applications. Examples are shown in Figure 2.27.

Some of these, like ANS and ethidium bromide, will bind non-covalently to particular regions of proteins and nucleic acids, with large changes in their fluorescent properties. ANS tends to bind to hydrophobic patches on proteins and partially unfolded polypeptides, with a blue shift and increase in fluorescence intensity. Ethidium bromide molecules intercalate between the base pairs of double-stranded DNA, resulting in a large increase in fluorescence which is used routinely to detect and visualize, for example, bands of nucleic acids in gel electrophoresis.

Some fluorescent dyes can be used to detect the presence of amyloid aggregates such as those found in BSE ('mad cow disease'), Alzheimer's and other protein misfolding diseases. Thioflavin T, for example, fluoresces strongly when bound to the β-sheet polypeptide structures in aggregated protein.

Other fluorescent probes (*e.g.* the dansyl group) are often used to covalently label an appropriate ligand, protein or other macromolecule. Such labels can then be used as reporter groups to detect binding, conformational changes or other effects (see, for example, Section 5.5).

### 2.4.6  Fluorescence Resonance Energy Transfer

Another kind of fluorescence quenching can occur when the fluorescence emission spectrum of the excited molecule overlaps with

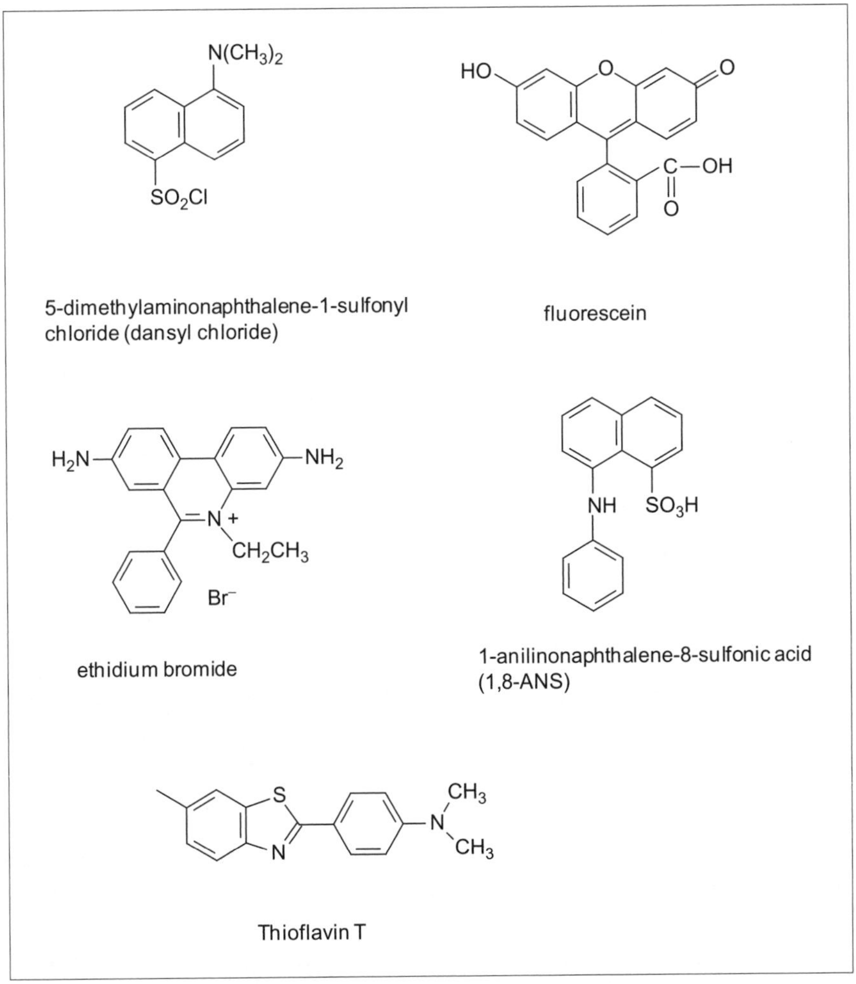

**Figure 2.27** Some common fluorescence probe molecules.

the electronic absorbance of another nearby group (Figure 2.28). Depending on the distance and relative orientation of the two groups, this can give rise to direct energy transfer from the excited 'donor' molecule to electronically excite the 'acceptor', with the donor molecule returning to the ground state without photon emission. This can either simply quench the fluorescence emission or, if the acceptor molecule is itself a fluorophore, give secondary fluorescence at a $\lambda_{max}$ corresponding to the acceptor. This effect can be used to identify nearby groups in macromolecular systems, or even to measure the separation between donor and acceptor groups in favourable cases, since the quenching efficiency ($E$) is a strong function of distance ($R$).

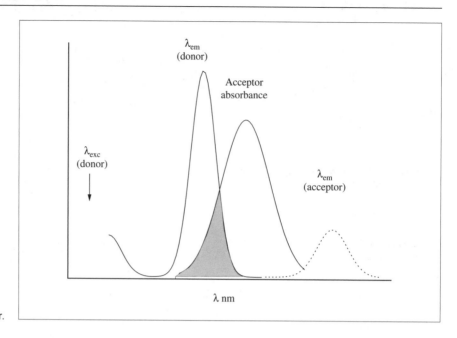

**Figure 2.28** Spectral overlap and fluorescence energy transfer.

The **fluorescence energy transfer** (FRET) quenching efficiency is defined as:

$$E = 1 - \phi_{DA}/\phi_D$$

where $\phi_{DA}$ and $\phi_D$ are the (donor) fluorescence intensities measured in the presence and absence of acceptor, respectively, under otherwise identical conditions.

The Förster mechanism of resonance energy transfer describes the effect in terms of dipole-induced dipole interactions between the donor and acceptor molecules, leading to an inverse 6th power ($1/R^6$) dependence:

$$E = R_0^6/(R_0^6 + R^6)$$

where $R_0$ is a constant for a particular donor/acceptor pair, giving the separation distance at which quenching is 50%. The value of $R_0$ depends on a number of factors including the quantum yield of the donor, the spectral overlap between the donor fluorescence and acceptor absorbance spectra (shown as the shaded area in Figure 2.28), and the relative orientations of the donor/acceptor groups. Typical $R_0$ values are in the range 2–6 nm.

One major technical challenge associated with this technique is how to introduce the donor and acceptor groups without perturbing the system too much.

---

**Worked Problem 2.6**

**Q:** Wu and Stryer[7] used FRET to estimate the molecular dimensions of the visual pigment protein, rhodopsin. They attached various fluorescent donor molecules at specific sites on the protein and used the intrinsic retinal chromophore as the acceptor group. For one particular fluorescent donor, attached to site B on the protein, $R_0$ was 5.2 nm and the observed fluorescence quenching was 36%. What is the distance between site B and the retinal chromophore?

**A:** Rearranging $E = R_0^6/(R_0^6 + R^6)$ gives $R^6/R_0^6 = 1/E - 1 = 1.78$ ($E = 0.36$)

$$R = R_0 \times (1.78)^{1/6} = 5.7 \, \text{nm}$$

### 2.4.7 Fluorescence Depolarization

If a fluorescent sample is excited using linearly polarized light, then predominantly only those molecules with transition dipoles oriented in the direction of polarization will be excited. Similarly, when the molecule fluoresces, the polarization of the emitted photon reflects the orientation of the transition dipole. Consequently, if the molecule is tumbling or rotating, and changes its orientation during the lifetime of the excited state, then the polarization of the emitted light will be different from that of the exciting beam. This is known as **fluorescence depolarization**. Since the effect depends on how fast the fluorescent group is tumbling, measurements of fluorescence depolarization can be used to get information about the dynamics of macromolecules in different environments. (See Chapter 4 for details about rotational diffusion and rotational relaxation times.)

### 2.4.8 Fluorescence Recovery after Photobleaching

Many fluorophores are sensitive to light and can be bleached if the irradiation is too intense. This can be an experimental inconvenience. However, it can also be turned to advantage by providing a method for following the diffusion of molecules on a microscopic scale. This is the basis of **fluorescence recovery after photobleaching** (FRAP).

Imagine a protein in a cell membrane that has been chemically labelled with a suitable fluorescent probe. Observation in the fluorescence microscope will show the distribution of this protein in the membrane. Now bleach a small area of the membrane using an intense flash of (laser) light. The photobleached molecules will no longer fluoresce but, if they and the surrounding molecules are free to move, the bleached area will gradually recover by molecular diffusion in the

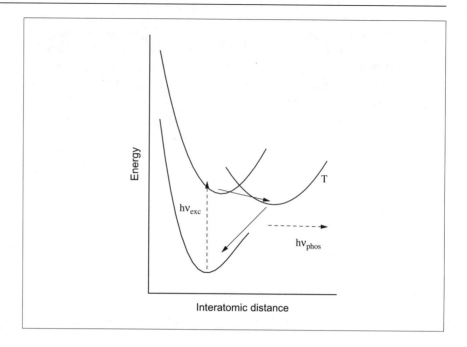

**Figure 2.29**  Energy diagram illustrating intersystem crossing.

membrane. The kinetics of this recovery process can give information about the lateral mobility of macromolecules and other components of the membrane.

For two-dimensional diffusion, the average distance travelled by a molecule in a membrane in time, $t$, is given by: $<x> = (4Dt)^{1/2}$, where D is the diffusion coefficient.

### 2.4.9  Phosphorescence and Luminescence

In some molecules, electrons in the singlet excited state may undergo a process known as 'intersystem crossing' to form a lower energy triplet state in which the spin state of the electron becomes reversed (Figure 2.29).

For such systems, return to ground state is formally forbidden by the Pauli exclusion principle and the lifetime of the triplet excited state is much longer than the more common singlet state. The return to ground state (when it eventually occurs) can also be accompanied by the emission of a photon. This is the phenomenon known as phosphorescence, in which objects can appear to glow in the dark for long periods of time.

The triplet state can be induced by chemical processes in the absence of light, and this is the basis for the phosphorescence or chemi-luminescence of many marine organisms.

The light emitted by courting fireflies is chemiluminescence generated by the luciferase enzyme-catalyzed oxidation of an organic substrate, luciferin, using ATP as energy source.

## 2.4.10 Fluorescence and Confocal Microscopy

One particularly useful application of fluorescence has been in the area of microscopy (see Chapter 8). Specific labelling of proteins or other components with fluorescent probes allows their position to be visualized within biological cells or tissues. Using a sharply focused laser beam as the exciting light, it is possible to observe fluorescence emission from tiny volume elements within the sample. By scanning of the laser focus, both up and down and side-to-side, observing the emission intensity at each point, a three-dimensional image of the fluorophore distribution within the sample can be built up.

Even greater depth penetration and resolution can be obtained by using two-photon excitation (Figure 2.30). Given a high enough light intensity, it is possible to achieve electronic excitation of the fluorescent group by simultaneous absorption of two photons (each with half the required energy) rather than the more usual single photon process. The subsequent fluorescence will take place regardless of how the excited state was produced.

This has two main advantages here. First, since very high light intensities are required for two-photon effects, emission will occur predominantly from the focal point of the laser beam where it is most intense, and much less from other regions where the beam is less focused. Secondly, the longer wavelengths used for two-photon excitation (usually in the near IR region) allow much greater depth penetration with less scattered background light than is normally found at the shorter wavelengths used for conventional fluorescence techniques.

## 2.5 Vibrational Spectroscopy: IR and Raman

Infrared spectroscopy probes the characteristic vibrational bands of chemical groups as the atoms move with respect to one another in

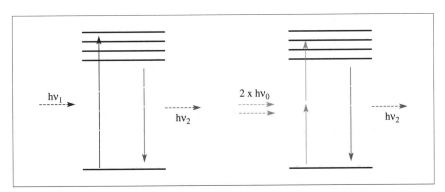

**Figure 2.30** One-photon excitation (left) and two-photon excitation (right), $v_0 = v_1/2$.

response to an oscillating electromagnetic field of the appropriate frequency. The normal modes of a particular molecule or group depend on the molecular structure, interatomic forces (bond strengths), and masses of the atoms concerned, but they will only be IR active if the vibrational mode involves a change in dipole moment.

Conventional IR spectrometers follow the same principles as described for UV/vis spectrometers but, bearing in mind that IR radiation is basically radiant heat, the detectors are usually sensitive thermocouples and the monochromator/spectrometer optics must be adapted to transmit in this spectral region. Samples are frequently prepared as thin films or solid dispersions between IR transparent optical surfaces (KBr discs, *etc.*).

Despite the utility of IR spectroscopy in other parts of chemistry, its application to biomolecules has been relatively limited. This is mainly due to sampling problems. Biomolecules are usually best studied in aqueous solution and are rarely soluble in the solvents normally required for IR. Water absorbs very strongly in the infrared and, apart from narrow window in the region of $1700\,cm^{-1}$, most of the spectral range is inaccessible even with highly concentrated biomolecule solutions. The accessible range can be extended a little by using deuterated water ($D_2O$) as solvent. Some improvements have come about in recent years with the development of Fourier transform (FTIR) instruments and multiple internal reflection (MIR) sampling methods.

The advantage of FTIR is that multiple spectral scans over a wide range can be accumulated to give a good signal-to-noise ratio even with strongly absorbing samples, and the background water spectrum can be measured separately and reliably subtracted to give the underlying sample spectrum. This is even further improved by placing the sample in an MIR device in which, rather than pass through the sample, the IR beam is reflected off the under surface of a crystal in contact with the sample. The IR beam penetrates only a very short distance into the sample at the crystal : liquid interface.

### 2.5.1  Raman Spectroscopy

The effect was first discovered by the Indian physicist, C. V. Raman, who was awarded the Nobel Prize for this in 1930.

Vibrational properties of molecules can also be studied using Raman spectroscopy, which is based on the **inelastic scattering** of light (Figure 2.31).

Note that the **scattering** of light by molecules is not to be confused with the absorbance or fluorescence described elsewhere. Light absorption requires light of specific wavelength or energies corresponding to particular electronic or vibrational transitions. Scattering, on the other hand, occurs at any wavelength and is just a general property of matter depending on electrical **polarizability** ($\alpha$).

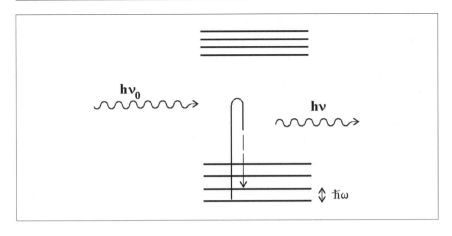

**Figure 2.31** Inelastic (Raman) scattering.

Imagine a molecule placed in an (oscillating) electromagnetic field. The electric field ($E$) at any one time will induce an electric dipole in the molecule, $\mu = \alpha E$. This oscillating electric dipole will itself act as a little radio antenna and will radiate in all directions at the frequency of the imposed field, $E$. If the imposed field is due to light, then the molecule will scatter a fraction of that light beam, without change of frequency. This is the basis for **elastic** or **Rayleigh scattering** of light. In addition, if the electrical polarization of the molecule involves stretching or bending of chemical bonds, then this might excite normal mode bending or stretching vibrations in the molecule itself. In this case, the induced frequency of dipole oscillation may be reduced by an amount corresponding to the normal mode frequency. The resulting frequency of the scattered light will be reduced accordingly because some of the light energy must be used to excite the molecular vibration. This shift in frequency (called the **Stokes shift**) is the basis for Raman spectroscopy.

The quantum picture of this process is perhaps easier to visualize. Most photons (particles of light) encountering a molecule will just pass on by unaffected. Occasionally a photon may bounce off the molecule without change of energy (Rayleigh/elastic scatter), or may set the molecule vibrating and bounce off with reduced energy (Stokes–Raman/inelastic scatter); like throwing stones at an empty oil drum, the stone may bounce off, leaving the drum vibrating at its resonant frequency. The process can be described as a **virtual transition** to a formally disallowed state (see Figure 2.31) with a rapid return to the ground state once it realises its mistake, so to speak. The virtual transition occurs effectively instantaneously, but can leave the molecule in a different vibrational level of the ground state from where it started.

Occasionally and especially at higher temperatures, the molecule may already be in a higher vibrational level and the scattered photon

The effect can be described by classical mechanics in terms of forced vibrations of harmonic oscillators. Here since the molecular polarizability, $\alpha$, changes slightly as the bond distorts, non-linear effects give rise to dipole oscillations at frequencies other than the imposed frequency. Raman himself seems to have been led to this discovery, at least in part, by his theoretical studies of the vibrations of musical instruments such as the violin.

The selection rules for IR and Raman are different, so some bands that are active in IR may be absent (or much weaker) in the Raman spectrum, and *vice versa*.

The population of vibrationally excited states is usually small at normal temperatures and decreases at higher frequencies. This is why anti-Stokes scattering is weak.

may instead pick up energy from this molecular vibration and appear at higher frequency (anti-Stokes shift, Figure 2.32).

Raman spectra are collected using an instrument somewhat like a spectrofluorimeter, but using a laser light source to give the intense monochromatic beam required (Figure 2.33).

Visible or near infra-red lasers are normally used and, since water is transparent in this region, this means that biological samples can be examined directly. It further helps that Raman scattering from water (unlike IR absorbance) is relatively weak. Examples of Raman spectra are shown in Figures 2.34 and 2.35.

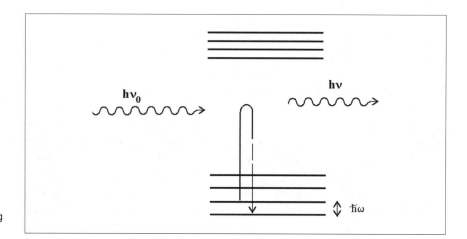

**Figure 2.32** Raman scattering with an anti-Stokes shift.

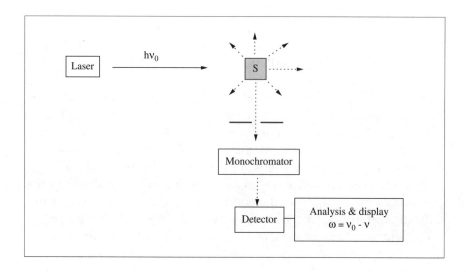

**Figure 2.33** Raman spectrometer.

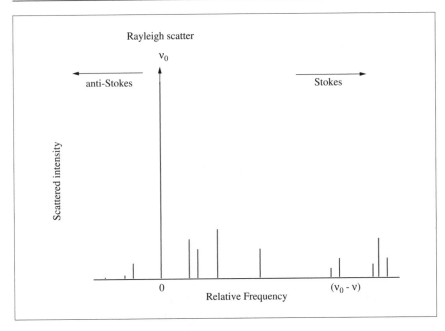

**Figure 2.34** A typical (idealized) Raman spectrum consisting of an intense Rayleigh line ($\nu_0$) and a series of lower frequency Raman bands corresponding to sample vibrational modes. The higher frequency anti-Stokes bands fall away rapidly and are rarely observed deliberately except for special purposes.

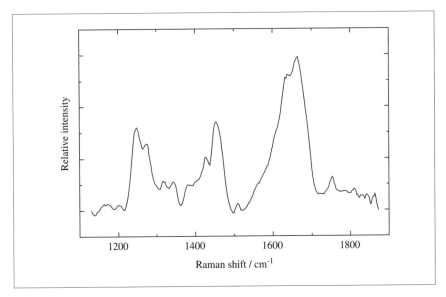

**Figure 2.35** Raman spectrum of a typical protein (collagen, 1 mg mL$^{-1}$) in water. (Note that most of the strong IR absorbing bands in water do not show up in the Raman spectra because of different selection rules.)

### Worked Problem 2.7

**Q:** What would be the wavelength of the Raman band corresponding to a carbonyl group frequency of 1650 cm$^{-1}$, assuming a laser wavelength of 546.0 nm?

$$\textbf{A:} \quad 546.0\,\text{nm} \equiv 18315\,\text{cm}^{-1} \pm 1650\,\text{cm}^{-1}$$

$$= 16665\,\text{cm}^{-1} \equiv 600.1\,\text{nm (Stokes)}$$

$$= 19965\,\text{cm}^{-1} \equiv 500.9\,\text{nm (anti-Stokes)}.$$

### 2.5.2  Resonance Raman, SERS and ROA

Raman scattering is inherently very weak and the complexity of over-lapping bands from biological macromolecules can sometime be hard to disentangle. However, if the wavelength of the laser light used for Raman excitation coincides with an electronic absorbance band in the sample, the Raman scattering from the chromophore can be greatly enhanced. This **resonance Raman** effect can be very useful in improving signals from dilute samples, or for selectively enhancing the Raman spectrum of just one component of a mixture. For example, this has been used to great advantage in studies of the visual pigment protein, rhodopsin, where the Raman spectrum coming from the retinal chromophore (which absorbs in the visible region) is intensified by resonance enhancement and stands out much more clearly from the background protein spectrum. UV lasers are now becoming available that permit direct resonance Raman studies of proteins and nucleic acids.

Another way to enhance Raman scattering is to adsorb the sample on to a roughened metal surface, or on to the surfaces of colloidal metal particles in solution. This is known as **surface enhanced Raman scattering** (SERS). The mechanism is not well-understood, but it is probably related to the strong local electromagnetic fields that occur when light shines on to the metal surface. When light encounters a metal it induces **surface plasmons** or intense collective motions of the conduction electrons in the metallic surface. This amplifies the oscillating electromagnetic fields close to the surface so that proteins or other molecules adsorbed on the surface experience much greater polarization and scatter more strongly.

In recent years a new technique called **Raman optical activity** (ROA) has been developed with specific applications to biological molecules in solution.[8] The polarizability and light scattering properties of chiral molecules will differ slightly depending on the polarization of the incident beam. This gives rise to small differences in the intensity of Raman scattering between left- and right-circularly polarized light. This can be used to generate an ROA spectrum that contains information about the chirality of the scattering groups as well as their normal mode frequencies.

## 2.6   NMR (Brief Overview)

Nuclear magnetic resonance spectroscopy (NMR) is based on the magnetic properties of certain atomic nuclei. It has become a cornerstone of modern chemical analysis and is now of major importance in studying the structure and dynamics of biological macromolecules. The technology involved is quite advanced and only the barest outline can be given here. A related text in this series gives more detail and a more rigorous treatment of this topic (see Further Reading).

Atomic nuclei containing an odd number of nucleons (protons or neutrons) can have a magnetic dipole moment ($\mu_B$) associated with their nuclear spin. When placed in a magnetic field ($B$) they act like tiny bar magnets (or compass needles) and tend to line up in the direction of the field. Nuclei of particular importance here are $^1H$, $^{13}C$, $^{15}N$ and $^{31}P$. These are all spin 1/2 nuclei which, according to quantum mechanics, can only exist in one of two possible states in a magnetic field: either parallel (low energy, $E = -\mu_B B$) or anti-parallel (high energy, $E = +\mu_B B$). The energy separation between these two states is:

$$\Delta E = 2\mu_B B$$

where $B$ is the strength of the magnetic field experienced by the nucleus, and $\mu_B$ is the component of the nuclear magnetic moment along the field axis.

The magnitude of the nuclear magnetic moment is related to a quantity known as the **gyromagnetic ratio** ($\gamma$) (see Table 2.2):

$$\mu_B = 1/2\,\hbar\gamma \quad (\hbar = h/2\pi)$$

Normally, of course, the majority of the nuclei will be in the lower energy state (see Figure 2.36). But, if an oscillating electromagnetic field is applied at the appropriate frequency, the nucleus may switch between the state and absorb energy from the applied field.

*'You know, what these people do is really very clever. They put little spies into the molecules and send radio signals to them, and they have to radio back what they are seeing.' . . . how Niels Bohr described NMR to Felix Bloch—one of the pioneers of this technique.*

**Table 2.2**  Gyromagnetic ratios and natural abundances of some nuclei

|  | $\gamma/10^7\,T^{-1}s^{-1}$ | % natural abundance |
|---|---|---|
| $^1H$ | 26.75 | 99.985 |
| $^{13}C$ | 6.73 | 1.11 |
| $^{15}N$ | −2.71 | 0.37 |
| $^{17}O$ | −3.63 | 0.037 |
| $^{19}F$ | 25.18 | 100.0 |
| $^{31}P$ | 10.84 | 100.0 |

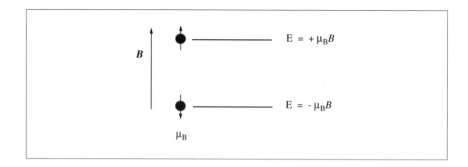

**Figure 2.36** Energy levels of nuclear magnetic dipoles in a magnetic field, $B$.

The resonant frequency is related to the energy separation:

$$h\nu = \Delta E = 2\mu_B B = \hbar\gamma B$$

so that the resonant frequency, $\nu = \gamma B/2\pi$

---

### Worked Problem 2.8

**Q**: What is the resonant frequency for a proton in a magnetic field of 14.1 Tesla?

**A**: For the proton ($^1H$ nucleus), $\gamma = 2.675 \times 10^8 \, T^{-1} s^{-1}$

$$\nu = \gamma B/2\pi = 2.675 \times 108 \times 14.1/2\pi$$
$$= 6.0 \times 108 \, Hz$$
$$= 600 \, MHz$$

---

This is the basis of NMR spectroscopy. Samples are placed in a strong magnetic field, usually between the poles of a superconducting magnet (Figure 2.37), and the resonant frequencies for each of the nuclei determined using an applied radio frequency (RF) electromagnetic field (usually in the 100 MHz to GHz range). The observed resonant frequency ($\nu$) for any particular atomic nucleus depends on two factors: the nature of the nucleus ($\mu_B$) and the magnetic field ($B$) that it sees. It is this second factor that makes NMR such a powerful tool, since $B$ is made up not just of the external applied field from the magnet ($B_0$) but additional contributions ($\delta B$) from other nuclei and chemical groups. This is because nearby atoms and groups (themselves acting like small magnets) will screen or shift the local magnetic field experienced by the nucleus. Consequently, NMR resonant frequencies are extremely sensitive to the chemical environment of the particular

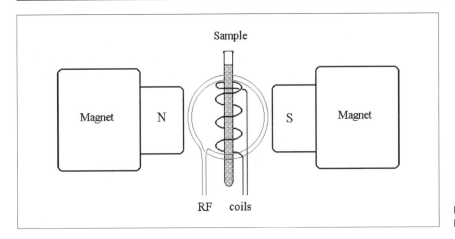

**Figure 2.37** Sketch of a typical NMR spectrometer.

atomic nucleus, and this can be used to give accurate structural information. The **chemical shift** and **spin–spin coupling** (**J-coupling** or **fine structure**) are ways of describing these additional contributions ($\delta B$). The characteristic patterns for specific chemical groups and different nuclei are well described in other books (see Further Reading).

In a basic NMR spectrometer, the resonant frequencies are obtained by scanning the applied magnetic field and detecting the values at which the oscillating field is absorbed. This is known as the continuous wave (CW) approach. It has the disadvantage that only one frequency is examined at a time and scanning across the entire spectrum is time-consuming. In order to overcome this disadvantage, the more powerful modern instruments use a somewhat different approach called **pulse** or **Fourier transform** FT-NMR. Here the applied magnetic field is kept constant and the magnetic nuclei in the sample are perturbed by applying a short (1–20 µs), broad-frequency microwave pulse that excites a range of different nuclei at once. The way in which the perturbed nuclear spins relax back to their equilibrium positions—called **free induction decay** (FID)—can be followed using **field coils** or **probes** placed close to the sample in the NMR spectrometer. The rate of decay of any particular nucleus to its equilibrium orientation is related to its resonant frequency, $v$, so the FID signal consists of a superposition of exponentially decaying oscillations made up of components arising from every perturbed nucleus in the sample. This can be analysed mathematically using Fourier transform methods to give the entire spectrum very rapidly. Since each pulse contains information about all nuclei simultaneously, data collection is very efficient and data from multiple pulses can be accumulated to give much greater sensitivity and improved signal-to-noise ratios.

Proton NMR is most common because of hydrogen's natural abundance and large gyromagnetic ratio. Studies with other nuclei such as $^{13}$C and $^{15}$N often require isotopic enrichment of the samples.

NMR spectrometers are often classified according to their resonant frequency for protons. For example, a '600 MHz' NMR instrument would comprise a 14.1 Tesla magnet in which $^{1}$H nuclei would resonate at roughly this frequency. Other nuclei would have different frequencies.

The way in which a perturbed nucleus relaxes to its equilibrium state in a magnetic field depends on a number of factors. An isolated nucleus cannot reorient by itself since the process requires transfer of spin angular momentum and NMR relaxation times depend on interactions of the perturbed nucleus with its surroundings. The **spin-lattice relaxation time** $(T_1)$ is related to the fluctuations in local magnetic field perceived by the nucleus as a consequence of other nearby nuclei, either in the same molecule or in surrounding solvent. This means that $T_1$ values depend on how fast the molecule or group containing the nucleus is tumbling or rotating in the sample. Measurements of NMR relaxation times can give important information about dynamic process in molecules such as rotational diffusion, and the rates of conformational fluctuations and other internal motions within macromolecules.

The **spin–spin relaxation time** $(T_2)$ is related to exchange of spin between equivalent nuclei. This is essentially a quantum mechanical effect that does not lead to loss of energy to the surroundings, but merely transfer of the magnetization to a nearby nucleus. Because of the Heisenberg uncertainty principle, this leads to a broadening of the spectral line, so that the width $(\Delta v)$ is roughly equal to $1/\pi T_2$. Spin–spin relaxation tends to be more efficient ($T_2$ is smaller) for slowly tumbling molecules and leads to significant broadening and overlap of NMR lines if rotational diffusion is slow on an NMR timescale. This is of particular significance when studying large proteins or other macromolecules in solution.

Spin-lattice anisotropy effects are even more significant with solid samples or biological membranes, where solution tumbling is very slow or non-existent. In such cases, techniques of **solid-state** NMR involving **magic-angle spinning** and other specialist methods have been developed.

The **uncertainty principle** is a universal consequence of quantum theory and the wave-like properties of matter, first expressed by Werner Heisenberg in the 1920s. In the form used here ($\Delta E.\Delta t \geq h/2\pi$), it states that the greater the uncertainty in the lifetime of a state ($\Delta t$) the narrower is the uncertainty in the energy or linewidth of that state ($\Delta E$), and *vice versa*. It applies generally to any system, not just NMR.

### Worked Problem 2.9

**Q**. What is the rotational diffusion (tumbling) time for a rigid spherical macromolecule of radius 1.5 nm in water at 20 °C? How does this compare to a typical NMR timescale? (The viscosity of pure water at 20 °C, $\eta = 1.002 \times 10^{-3}\,\mathrm{N\,s\,m^{-2}}$)

**A**: Using the methods described in Chapter 4 (Section 4.5):

$$\tau_{rot} = 8\pi\eta R^3/2kT$$
$$= 8\pi \times 1.002 \times 10^{-3} \times (1.5 \times 10^{-9})^3/(2 \times 1.381 \times 10^{-23} \times 293)$$
$$= 1 \times 10^{-8}\mathrm{s}(10\,\mathrm{ns})$$

$$1/\tau_{rot} \equiv 100\,\mathrm{MHz}(\text{comparable to typical NMR frequencies}).$$

Spin–spin interactions can affect the intensities of NMR spectral lines in addition to affecting the chemical shift and relaxation properties of a given nucleus. One example of this is the **nuclear Overhauser effect** (NOE) in which the magnetization state of one nucleus can affect the spin population, and hence the NMR absorbance, of another nearby nucleus. The magnitude of the effect depends on a dipole–dipole interaction between the two nuclei, and is therefore very sensitive to the distance between the nuclei, with a $1/r^6$ dependence. This makes it possible to measure distances between specific atoms and forms the basis for determining three-dimensional molecular structures.

In addition to enhanced sensitivity, pulse FT-NMR techniques have given rise to numerous ingenious methods whereby, using different sequences of microwave pulses, different features of NMR spectra can be enhanced or suppressed in a controlled fashion. Such techniques are particularly powerful when combined with **multinuclear NMR**, involving other nuclei such as $^{13}$C and $^{15}$N in addition to $^1$H, often requiring isotopically enriched samples. These experiments give rise to **multidimensional** NMR spectra in which the chemical shifts and intensities of NMR lines for different nuclei are plotted along different axes. Correlations and interactions between different nuclei show up more clearly in such representations, and a large number of acronyms are used to designate different pulse sequences and methods, for example:

- COSY (correlation spectroscopy);
- TOCSY (total correlation spectroscopy);
- NOESY (nuclear Overhauser effect spectroscopy);
- HSQC (heteronuclear single quantum coherence).

See Figure 2.38 for an example.

One of the more important current applications of high-resolution multinuclear NMR is in the determination of the three-dimensional structures of proteins and other macromolecules in solution. These techniques are complementary to crystallographic methods (see Chapter 8). A typical experimental procedure is described in Box 2.3.

> The nuclear Overhauser effect is quite short range because of the $1/r^6$ dependence. Longer range distance, orientation and molecular dynamics information can be obtained from measurements of **residual dipolar coupling** effects with partially oriented samples.[9]

---

### Box 2.3 Stages in a typical protein structure determination by NMR

---

1. *Sample preparation*: Sample concentrations in the range 0.5–2 mM (several mg cm$^{-3}$) are usually required and samples

must be enriched in other NMR nuclei ($^{13}C$, $^{15}N$). This can be achieved by recombinant DNA methods in isotopically enriched media.

2. *Data acquisition and processing*: Collect NMR data using high resolution (500MHz or more) spectrometers, using different pulse sequences to selectively stimulate specific nuclei.

It is now possible to determine the structure of proteins within living cells using NMR techniques (see ref. 10).

3. *Sequence-specific assignment*: Each peak in the NMR spectrum (of which there will be thousands) corresponds to a particular atom (H, C, N, *etc.*) in the structure. These can be assigned to specific backbone or side chain residues in the sequence using chemical shift data and correlation methods (COSY, TOCSY, HSQC, *etc.*). Intra- and inter-residue

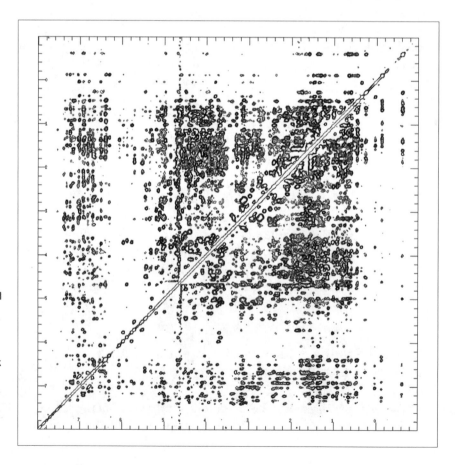

**Figure 2.38** An example of a two-dimensional NOESY spectrum of a DNA fragment in $D_2O$ (courtesy of Dr Brian Smith). This is a two-dimensional contour map showing how efficiently nuclear magnetization can be transferred from one $^1H$ nucleus to another nearby in the structure. Each off-diagonal peak corresponds to a pair of protons and the size of the peak can be related to the distance between them (the bigger the peak, the closer they are together in the molecule).

'through bond' assignments can be linked to create a complete protein shift map.

4. *NOE cross-peak identification and assignment*: Use NOESY to determine pairwise distances between specific nuclei in the structure, identifying in particular those atoms that are close together in the structure. These data provide a list of structural restraints for subsequent model building.

5. *Structure calculation, model building and refinement*: The large number of pairwise structural restraints should be sufficient to define a topologically consistent three dimensional structure for the protein.

What usually results is a family of closely similar structures, all of which are consistent with the NOESY data. Within the limits of experimental error, this family of structures represents the dynamic conformational population of the protein in solution and quite often emphasizes the inherent flexibility of biological macromolecules. Figure 2.39

**Figure 2.39** Typical protein NMR structure showing a population of conformations consistent with the observed conformational restraints (courtesy of Dr Nicola Meenan).

### Summary of Key Points

1. Spectroscopy is based on the interaction of electromagnetic waves with matter. Different frequency ranges will probe different electronic and magnetic properties of molecules. UV/visible absorbance and fluorescence spectroscopy is based on electronic transitions. Raman spectroscopy gives information about vibrational bands in the IR region, while NMR uses radio frequency transitions of nuclear dipoles to give detailed structural information.
2. Biological chromophores have characteristic UV/visible absorbance and fluorescence spectra that can give general information about molecular environments.
3. Circular dichroism measurements can give information about secondary structure in proteins and nucleic acids.
4. Vibrational spectroscopy: conventional IR spectroscopy with biological molecules in water is difficult, but this can be overcome by using laser inelastic light scattering (Raman) techniques.
5. Multinuclear NMR methods are used to study the conformation and dynamics of biological macromolecules in solution.

### Problems

**2.1.** What is giving rise to the electromagnetic radiation coming from: (a) red hot charcoal; (b) a fluorescent light bulb; (c) a microwave oven; (d) an X-ray generator tube; (e) a laser; (f) a synchrotron; (g) a mobile phone?

**2.2.** Without using a calculator, what fraction (percentage) of light would pass through a sample with an absorbance (A) of: (a) 1; (b) 2; (c) 5; (d) 0?

**2.3.** (a) What is the absorbance corresponding to the following transmittances: 1%, 5%, 25%, 50%, 90%, 100%? (b) What fraction of light would be transmitted by samples having the following absorbances: 0.2, 0.5, 1, 1.5, 10?

**2.4.** A typical laboratory spectrophotometer is quoted as having a 'stray light rejection of 0.1%' What absorbance reading would this instrument give for samples with a (true) absorbance $A = 1.0$, 2.0 or 3.0?

**2.5.** Why is the 'stray light effect' more significant at high absorbance values?

**2.6.** (a) Estimate $\varepsilon_{280}$ for each of the following proteins: lysozyme (six Trp, three Tyr, three Phe); insulin (zero Trp, four Tyr, three Phe); ribonuclease (zero Trp, six Tyr, three Phe); human serum albumin (one Trp, 18 Tyr, 31 Phe). (b) In each of these, which amino acid (if any) makes the dominant contribution?

**2.7.** For a particular recombinant protein solution (produced by genetic engineering methods), the total concentration estimated by 280 nm absorbance measurements is $0.8 \, \text{mg mL}^{-1}$. However, biological activity measurements suggest that the concentration is only $0.6 \, \text{mg mL}^{-1}$. What are the possible reasons for this discrepancy?

**2.8.** What other methods might be used for measuring protein concentrations?

**2.9.** The experimental CD spectra of two proteins (called 'ABA-1' and 'RS') of unknown structure are shown below (Figure 2.40). What is the likely dominant secondary structure in each case?

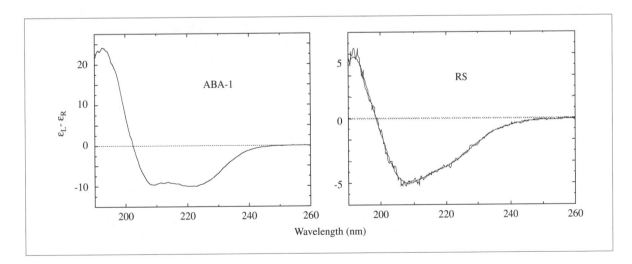

**Figure 2.40**

**2.10.** Mixtures of amino acids derived by acid hydrolysis of proteins are optically active, whilst mixtures of nucleotide bases are not. Why?

**2.11.** Some experiments attempting to simulate the origin of life by chemical synthesis from inorganic starting materials have yielded mixtures of amino acids (and other molecules) similar to those found in biology. But the solutions do not show any circular dichroism. Why?

**2.12.** Fluorescence emission spectra ($\lambda_{em}$) are usually independent of excitation wavelength ($\lambda_{exc}$). Why?

**2.13.** Suggest why $A < 0.2$ is recommended to minimize inner filter effects in fluorescence spectroscopy using a standard 1-cm path length cuvette. (Hint: calculate the % transmission.)

**2.14.** Typical molecular thermal energy is of order kT per degree of freedom (RT per mole). What vibrational frequency would this correspond to at room temperature?

**2.15.** The fluorescence spectrum of a protein (Figure 2.24) excited with 285 nm light shows a small Raman band around 318 nm coming from the solvent water. What vibrational band does this correspond to?

## References

1. S. C. Gill and P. H. Von Hippel, Calculation of protein extinction coefficients from amino-acid sequence data, *Anal. Biochem.*, 1989, **182**, 319–326.
2. M. Bradford, A rapid and sensitive method for the quantitation of microgram quantities of protein utilizing the principle of protein-dye binding, *Anal. Biochem.*, 1976, **72**, 248–254.
3. F. M. Pohl and T. M. Jovin, Salt-induced cooperative conformational change of a synthetic DNA. Equilibrium and kinetic studies with poly-(dG-dC), *J. Mol. Biol.*, 1972, **67**, 375–396.
4. S. M. Kelly, T. M. Jess and N. C. Price, How to study proteins by circular dichroism, *Biochim. Biophys. Acta*, 2005, **1751**, 119–139.
5. J. R. Lakowicz and G. Weber, Quenching of protein fluorescence by oxygen. Detection of structural fluctuations in proteins on the nanosecond timescale, *Biochemistry*, 1973, **12**, 4171–4179.
6. H. Y. Tsien, The green fluorescent protein, *Annu. Rev. Biochem.*, 1998, **67**, 509–544.

7. C. W. Wu and L. Stryer, Proximity relationships in rhodopsin, *Proc. Natl. Acad. Sci. U. S. A.*, 1972, **69**, 1104–1108.
8. L. D. Barron, A. Cooper, S. J. Ford, L. Hecht and Z. Q. Wen, Vibrational Raman optical-activity of enzymes, *Faraday Discuss.*, 1992, **93**, 259–268.
9. A. Bax and A. Grishaev, Weak alignment NMR: a hawk-eyed view of biomolecular structure, *Cur. Opin. Struct. Biol.*, 2005, **15**, 563–570.
10. D. Sakakibara, A. Sasakil, T. Ikeyal, J. Hamatsu, T. Hanashima, M. Mishima, M. Yoshimasu, N. Hayashi, T. Mikawa, M. Wälchli, B. O. Smith, M. Shirakawa, P. Güntert and Y. Ito, Protein structure determination in living cells by in-cell NMR spectroscopy, *Nature*, 2009, **458**, 102–105.

## Further Reading

J. Cavanagh, W. J. Fairbrother, A. G. Palmer III, M. Rance and N. J. Skelton, *Protein NMR Spectroscopy: Principles and Practice*, Academic Press, New York, 2nd edn, 2007.

G. Fasman (ed.), *Circular Dichroism and the Conformational Analysis of Biomolecules*, Plenum, New York, 1996.

J. M. Hollas, *Basic Atomic and Molecular Spectroscopy*, RSC Tutorial Chemistry Text, Royal Society of Chemistry, Cambridge, 2002.

P. Kelly and J. D. Woollins, *Multi-element NMR*, RSC Tutorial Chemistry Text, Royal Society of Chemistry, Cambridge, 2004.

J. R. Lakowicz, *Principles of Fluorescence Spectroscopy*, Plenum, New York, 1983.

P. R. Selvin, The renaissance of fluorescence energy transfer, *Nat. Struct. Biol.*, 2000, **7**, 730–734.

D. Sheehan, *Physical Biochemistry: Principles and Applications*, Wiley, New York, 2000, ch. 3.

I. Tinoco, K. Sauer, J. C. Wang and J. D. Puglisi, *Physical Chemistry: Principles and Applications in Biological Sciences*, Prentice Hall, Upper Saddle River, NJ, 2002, 4th edn, ch. 10.

K. E. van Holde, W. C. Johnson and P. S. Ho, *Principles of Physical Biochemistry*, Prentice Hall, Upper Saddle River, NJ, 1998, ch. 8–12.

# 3
# Mass Spectrometry

Mass spectrometry (MS for short) is rather different from the other kinds of spectroscopy we considered in the previous chapter. Instead of using electromagnetic radiation, MS separates and analyses mixtures of molecules—or molecular fragments—on the basis of their mass ($m$) and charge ($z$). This is done using a combination of electric and magnetic fields to control the flight paths of molecules in a vacuum chamber—in much the same way as beams of electrons are manipulated in a television picture tube. The results can give quite detailed information about molecular structure and interactions.

## Aims

In this chapter we explore the theoretical and experimental basis of various mass spectrometry techniques and their applications to biomolecules. By the end you should be able to:

- Explain how charged particles move under the influence of electric and magnetic fields
- Outline the basic components of a mass spectrometer
- Describe different ways in which particles may be separated according to their $m/z$ ratios
- Describe how molecular ions are produced experimentally
- Use MS data to determine relative molecular masses of proteins and other (macro)molecules

## 3.1    Introduction

Mass spectrometry is based on two fundamental physical properties of charged particles:

- Charged particles in an electric field ($E$) will move along the direction of the field.
- Charged particles moving through a magnetic field ($B$) will experience a force at right angles to their direction of movement.

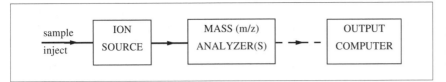

**Figure 3.1**  A typical mass spectrometer consists of a vacuum chamber fitted with a sample ionization source, electromagnetic focusing and acceleration electrodes to control the speed and trajectory of the ion beam, mass (m/z) analyser and detector.

Both these effects give rise to the separation of charged particles in terms of their mass-charge ratio (*m/z*) (see Figure 3.1).

J. J. Thomson's studies of the deflection of 'cathode rays' by electric and magnetic fields led to his discovery of the electron, for which he was awarded the Nobel Prize in 1906. He went on to obtain the first mass spectra of 'positive rays' of inert gases and other ions by similar techniques.

**Worked Problem 3.1**

**Q**: Why does a mass spectrometer require a high vacuum?

**A**: To avoid molecular collisions that would disrupt the ion flight path.

According to the kinetic theory of gases, the mean free path (*i.e.* the average distance travelled before colliding with another molecule) is given by:

$$l = kT/(2^{1/2} P \sigma)$$

where *P* is the gas pressure and $\sigma$ is the collision cross-section area for the colliding gas molecules. Typical collision cross-sections for small molecules are around $1\,nm^2$. For atmospheric pressure (1 atm, $1.013 \times 10^5\,Pa$) at 300 K, this gives a mean free path $l \approx 3 \times 10^{-8}\,m$ (30 nm, 300 Å), so molecules would collide after travelling just a few molecular diameters. For a typical unimpeded flight path in a mass spectrometer in excess of 1 metre, a pressure (P) inside the instrument of less than $3 \times 10^{-8}\,atm$ would be required.

## 3.2    Ion Sources

The experimental challenge for MS is how to get a stream of molecular ions into the high vacuum of the instrument in a suitable form for mass analysis. Sample ionization techniques have proven to be the key to successful applications of MS to biomolecular problems. Most compounds, especially high RMM and polar materials, have a very low vapour pressure and do not easily evaporate, even under the high vacuum in the MS. With small compounds, it is usually possible to produce volatile versions by careful **chemical derivatization** (*e.g.* permethylation, acylation, esterification), and this is the basis for most routine chemical applications of MS. But this is rarely satisfactory for larger biomolecules and alternative strategies have to be adopted.

## 3.3    Ionization Methods

### 3.3.1    Electron Impact and Chemical Ionization

These methods usually require chemical derivatization to produce volatile species with sufficient vapour pressure to form a gas in the MS sample compartment. The sample may be introduced either by itself, for **electron impact** (EI) methods, or mixed with a large excess of another gas for **chemical ionization** (CI). The volatile sample mixture is bombarded with a beam of electrons with energies (typically up to 100 eV) which may be captured to produce negatively charged species ($M^-$), or where electron impact is sufficient to displace electrons and produce positive ions ($M^+$). This may also result in fragmentation of the unstable radicals produced from the sample molecules, and the characteristic fragmentation patterns are frequently useful in structural identification.

Gentler ionization (CI) may be achieved by mixing the sample with a suitable carrier gas (*e.g.* methane or ammonia) to reduce sample fragmentation. Here the carrier gas takes the brunt of the electron bombardment, creating positively charged $CH_4^+$, $NH_3^+$ and other radicals which can then transfer positive charge ($H^+$) to the sample molecules by more gentle collisions. The resulting molecular ions ($MH^+$) are also intrinsically more stable than the radicals produced by direct electron impact, so molecular fragmentation is reduced. This simplifies the resulting mass spectrum and makes accurate mass determination easier, but reduces the often useful structural information that may be obtained by analysis of fragmentation patterns.

Both these ionization methods are best suited to relatively small molecules and have been used extensively for analysis of simple permethylated carbohydrates. They cannot usually be applied to much less volatile species such as peptides, proteins or other macromolecules.

### 3.3.2    Fast Atom Bombardment

In **fast atom bombardment** (FAB), the sample is introduced as a viscous liquid—often dissolved in a solvent such as glycerol (propane-1,2,3-triol)—which is then bombarded with a stream of energetic atoms or ions (typically argon or xenon at energies up to 30 keV) rather than electrons. This results in sputtering of molecules from the sample and a proportion of them are also ionized. Fragmentation may also occur during collision and ionization. This cloud of ions is then directed into the MS for analysis. This method is typically useful for peptides or small proteins of RMM up to about 10 000.

**Sputter**: to emit particles in an explosive manner.

### 3.3.3 Electrospray Ionization

In **electrospray ionization** (ESI), the sample is injected as a dilute solution through a fine orifice or capillary to produce a fine spray or stream of droplets. A voltage may be applied to the capillary and surrounding electrodes so that the droplets carry electrostatic charge. The solvent is driven off by a stream of dry gas in the sample chamber and, as the solvent evaporates, electrostatic repulsion breaks the droplets up further, leaving the charged particles (frequently single molecules) to be directed into the high vacuum chamber of the MS. This is a relatively gentle ionization method and is particularly useful for proteins and other large molecules. But although little or no fragmentation occurs under optimal conditions, each macromolecule does give rise to a number of peaks in the MS spectrum because of the range of charges that molecules may carry ($MH^+$, $MH_2^{2+}$, *etc.*). This has to be taken into account during interpretation of the spectra. The advantage, however, is that it does give a number of peaks from which more accurate mass data may be calculated. An example of this is given in Figure 3.5. Variants of this method can be used to sample directly the eluant from chromatographic separations.

### 3.3.4 Matrix-Assisted Laser Desorption Ionization

In this method, abbreviated MALDI, the sample is incorporated into a solid matrix made up of an ultraviolet (UV) or light-absorbing compound such as 2,5-dihydroxybenzoic acid. An intense pulse of laser light (typically around 390 nm) is focused onto the surface of this matrix, where the absorbed energy produces local heating and desorption of some of the matrix. These desorbed molecules, many of them thermally ionized, are then directed into the MS chamber. Again, this relatively gentle desorption technique is particularly useful for potentially fragile macromolecules. The pulse of ions produced by this method makes it particularly suited to **time-of-flight** (TOF) analysis methods (see Section 3.4.2).

Nowadays, the more sophisticated MS instruments combine a range of ionization and analysis methods.

## 3.4    Mass Analysers

Once the molecular ions are produced they are usually accelerated and focused using high voltage electric fields and suitable electrodes, prior to injection into the mass analyser. A number of different mass analysis methods are now in use and new techniques are continually being developed. However, all the methods described here have one thing in common: they all separate ions on the basis of their mass-to-charge ratio ($m/z$).

### 3.4.1  Magnetic Analysis Methods

Earlier versions of mass spectrometers were based on magnetic separation and this technique is still in use today (Figure 3.2).

A charged particle ($ze$) moving in a uniform magnetic field will tend to follow a circular trajectory in which the (inward) magnetic force ($zevB$) is balanced by the (outward) centrifugal force ($mv^2/r$):

$$mv^2/r = zevB \qquad (3.1)$$

so that:

$$r = mv/zeB \qquad (3.2)$$

The velocity ($v$) is determined by the voltage on the electrodes used to accelerate the ions. The kinetic energy gained by the particle passing through an electrostatic potential ($V$) is equal to $zeV$:

$$\frac{1}{2}mv^2 = zeV \qquad (3.3)$$

Rearrangement of eqn (3.1) and eqn (3.2) gives:

$$r = (2mV/zeB^2)^{1/2} \qquad (3.4)$$

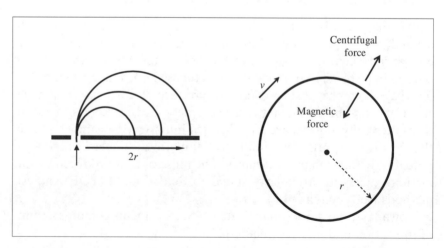

**Figure 3.2**  Charged particles moving in a magnetic field.

This shows how the radius of the ion trajectory or displacement of the particle in the magnetic field depends on $m/z$—since all other terms in eqn (3.4) are fixed.

### 3.4.2  Time-of-Flight Methods

Direct measurement of the velocity ($v$) of the charged fragments is used in time-of-flight (TOF) techniques. The kinetic energy gained by the particle passing through an electrostatic potential ($V$) is equal to $zeV$, so that:

$$\frac{1}{2}mv^2 = zeV \quad \text{or} \quad v = (2zeV/m)^{1/2} \tag{3.5}$$

The time-of-flight of the particle along a fixed path will depend on the inverse of its velocity:

$$\text{TOF} \propto 1/v = (m/z)^{1/2}(1/2eV)^{1/2} \tag{3.6}$$

For this kind of analyser, a discrete pulse of molecular ions is injected into a field-free 'drift tube', measuring the time taken for ions to reach the detector (see Figure 3.3). This method of analysis is particularly suited to MALDI ionization methods and can be used for quite large macromolecules. As before, the separation of molecular ions is on the basis of their mass/charge ratio ($m/z$).

### 3.4.3  Quadrupole Analysers

Quadrupole separation relies on the movement of ions in a non-uniform electric field. The beam of ions is directed into the non-uniform electric ('quadrupole') field formed in the cavity between four parallel cylindrical electrodes (Figure 3.4).

An oscillating potential of $\pm[V_0 + V.\cos(\omega t)]$ is applied to these electrodes, causing the molecular ions to follow complicated motion, most of which will result in collision with the electrodes or the cavity wall. For a given combination of voltages ($V$, $V_0$) and frequency ($\omega$), only ions with particular $m/z$ will pass through to hit the detector. Typical values are: $V_0 = 500$–$2000$ volts; $V = 0$–$3000$ volts; and $\omega = 1$–$2$ MHz.

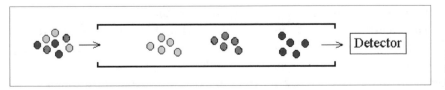

**Figure 3.3**  Time-of-flight separation of particles with different velocities in a drift tube.

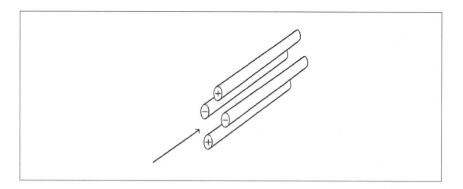

**Figure 3.4**   Quadrupole analyser.

### 3.4.4  Ion Trap and Ion Cyclotron Resonance FT-MS Methods

A number of more sophisticated variants of quadrupole and other analysers based on complex ion optics have been developed in recent years.

**Ion trap** methods use a combination of electrodes to form electrostatic fields within which ions are trapped in specific orbits depending on $m/z$ and the applied field. Ions with specific $m/z$ ratio may be detected by varying this applied voltage.

**Ion cyclotron resonance** is related to quadrupole techniques, but with the addition of a strong magnetic field, $B$. This induces the ions into (roughly) circular orbits with angular frequency:

$$\omega = v/r = zeB/m \tag{3.7}$$

Application of an oscillating electric field of the same frequency, $\omega$, leads to cyclotron resonance in which the ions absorb energy from the electric field and move increasingly faster, with increased orbital radius, whilst keeping $v/r$ constant. Ions of particular $m/z$ are therefore selected on the basis of the frequency of the applied field. This makes this method particularly appropriate for frequency-domain Fourier Transform (FT) analysis, and gives potentially very high resolution.

## 3.5    Detection

Early mass spectrometers used photographic plates to detect the positions of molecular ions. Nowadays these have been replaced with more sensitive and convenient electronic detectors such as photomultiplier tubes in which ion collision with the photomultiplier plate ejects a pulse of free electrons (just as in the detection of photons). This electron pulse is then accelerated through a cascade of electrodes where collision at each stage amplifies the number of electrons in the pulse (hence the term 'photomultiplier') to give a measurable electronic signal (see Figure 2.9 for a diagram of a photomultiplier.)

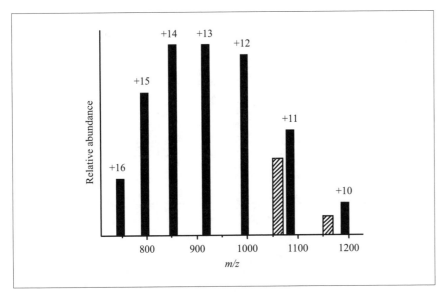

**Figure 3.5** Example mass spectrometry data for an intact protein of theoretical RMM 11924.2 (solid bars) with calculated charge (*z*) indicated. The shaded peaks indicate possible impurities.

## 3.6    Applications of MS

Because mass spectrometry can determine RMMs with high precision, it has found numerous applications in areas where other techniques can give only vague answers. The following gives some practical examples.

### 3.6.1  Relative Mass of Proteins

 Figure 3.5 shows part of a typical ESI-MS profile for a pure protein (*m/z* values are given in Table 3.1). Each peak in the spectrum corresponds to a different mass-to-charge ratio produced on the molecular ions during the electrospray process. Adjacent peaks will usually differ in both charge and mass by one unit. How can we interpret such data?

**Reminder:** $m = M + z$, where $M$ is the relative mass of the neutral protein molecule.

> **Worked Problem 3.2**
>
> **Q**: Use the mass spectrometry data from Table 3.1 to determine the RMM for the protein.
>
> **A**: Select any peak. This will have the value *m/z*, but we do not yet know either *m* or *z* separately. Its neighbouring peak will correspond to the molecular ion with a mass-to-charge ratio

**Table 3.1**  Data for Figure 3.5

| Measured | | Calculated | | |
| --- | --- | --- | --- | --- |
| m/z | Δ | z | m | M = m − z |
| 1193.417 | 108.381 | 10.0021 | 11936.66 | 11926.66 |
| 1085.036 | 90.353 | 10.9978 | 11932.99 | 11922.00 |
| 994.683 | 76.437 | 12.0000 | 11936.22 | 11924.22 |
| 918.246 | 65.517 | 13.0001 | 11937.31 | 11924.31 |
| 852.729 | 56.782 | 14.0000 | 11938.19 | 11924.19 |
| 795.947 | 49.685 | 14.9997 | 11939.00 | 11924.00 |
| 746.262 | – | | | |
| | | | Mean | 11924.23 |
| | | | s.d. | 1.48 |

$(m+1)/(z+1)$, so the difference between them ($\Delta$) will be given by:

$$\Delta = (m/z) - (m+1)/(z+1) = (m/z - 1)/(z+1)$$

This may be rearranged to give:

$$z = (m/z - 1)/\Delta - 1$$

This gives us $z$ for the chosen peak, so we can then calculate $m$ for that peak. The mass of the (neutral) parent species, $M = m - z$.

The results of such a calculation (Table 3.1) give M = 11924.23 ($\pm 1.48$) compared to the theoretical value of 11924.2 (calculated from the amino acid composition).

It is important to use a series of peaks in such data in order to get consistent results. For example, the shaded peaks in Figure 3.5 (with $m/z$ of 1060.1 and 1160.3) would not fit into a consistent pattern with the other data. They probably arise from impurities or degradation products.

### 3.6.2 'Ladder' Sequencing

Mass spectrometry may be used to determine primary sequences. Partial sequential chemical or enzymic degradation of peptides or proteins can yield a mixture of peptides with progressively shorter sequences. For example, consider the following (hypothetical)

**Table 3.2**  Example of 'ladder' of fragments

| Polypeptide sequence | Mass/AMU | Δ | Amino acid |
|---|---|---|---|
| X-Ser-Gly-Trp-Glu-Asp-Leu-Ile-Lys-Met | 10531 | 131 | Met |
| X-Ser-Gly-Trp-Glu-Asp-Leu-Ile-Lys | 10400 | 128 | Lys/Gln? |
| X-Ser-Gly-Trp-Glu-Asp-Leu-Ile | 10272 | 113 | Leu/Ile? |
| X-Ser-Gly-Trp-Glu-Asp-Leu | 10159 | 113 | Leu/Ile? |
| X-Ser-Gly-Trp-Glu-Asp | 10046 | 115 | Asp |
| X-Ser-Gly-Trp-Glu | 9931 | 129 | Glu |
| X-Ser-Gly-Trp | 9802 | 186 | Trp |
| X-Ser-Gly | 9616 | 57 | Gly |
| X-Ser | 9559 | 87 | Ser |
| X | 9472 | | |

polypeptide sequence that has been partially digested with a proteolytic enzyme (carboxypeptidase) to yield the 'ladder' of fragments shown in Table 3.2.

Mass spectrometry of the mixture gives a series of peaks of decreasing mass, each with a mass difference (Δ) corresponding to the loss of one amino acid from the *C*-terminus. Since the mass of each amino acid residue is known accurately, one can work backwards from this to construct the partial amino acid sequence of the original polypeptide. This is known as ladder sequencing, and similar approaches can be used for sequencing DNA and polysaccharides.

Note, however, that the method is not entirely unambiguous because some amino acids have the same residue mass. For example in this case, it is not possible to distinguish lysine from glutamine (Δ = 128) or leucine from isoleucine (Δ = 113) by MS alone.

In practice, it is rarely possible to sequence lengths of more than 20 or so amino acids by these methods because the mixtures get too complex and degradation patterns are uneven. However, a range of short sequences from a collection of different peptides from the one protein ('peptide mapping') is often sufficient to identify the protein. If the different peptide sequences overlap, then the entire protein sequence can be determined.

### 3.6.3  Proteomics

Now that human (and other) genome sequences are available, it is important to be able to identify the proteins associated with each of the genes and their interactions. This is generally known as 'proteomics'. A biological cell contains thousands of different proteins and mass spectrometry can be used to identify them. Since different

'Proteomics is the study of the function of all expressed proteins.'[1]

proteins usually differ in size and charge, they can be separated using chromatographic or electrophoresis methods (see Chapter 7). For example, two-dimensional gel electrophoresis of the proteins from disrupted cells gives 1000 or more separate spots, each corresponding to a different protein. Using the MS techniques described above, it is possible to analyse the amino acid sequence of each of these proteins and, since we know the genetic code, identify the genes from which they are derived. The advantage of mass spectrometry here is that it is very sensitive and requires only very small amounts of sample.

Peptide 'mass fingerprinting' can also be used to identify proteins after treatment with specific proteases (Figure 3.6). For example, trypsin is a proteolytic enzyme (protease) that specifically cleaves Arg-Lys peptide bonds. Consequently, trypsin digestion will give a range of peptides corresponding to cleavage only at such sites and the MS pattern of the peptide mixture (the mass fingerprint) that will be specific for a given protein sequence. This fingerprint can then be compared to a database of theoretical peptide fingerprints (predicted from genome sequence data) to identify the protein.

Proteins (or other molecules) that interact strongly with each other in the cell may become tightly associated and co-purify in cell extracts. Mass spectrometry of purified complexes can help identify

A **protease** is any enzyme that catalyses the hydrolytic cleavage (**proteolysis**) of peptide bonds in proteins. Pepsin and trypsin are typical proteases found in our digestive system.

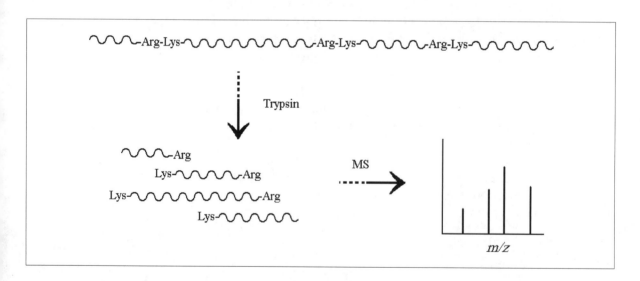

**Figure 3.6** Peptide mass fingerprinting. Trypsin cleaves only at adjacent Arg-Lys residues to give a mixture of peptides with characteristic *m/z* pattern.

the interacting partners. Techniques are now also available to study more weakly interacting proteins and complexes that might otherwise be torn apart by dehydration and other disruptive forces in the mass spectrometer.[2]

---

### Summary of Key Points

1. Mass spectrometry separates molecular ions on the basis of mass-to-charge ratio ($m/z$).
2. A range of gentle ionization methods and mass analysers are now available to study biological macromolecules.
3. Primary sequence and interaction information can be obtained.

---

### Problems

**3.1.** In a mass spectrum of an intact protein, why do adjacent peaks differ in mass by 1 amu? What protein groups most likely carry these charges?

**3.2.** How fast might the following ions be travelling after acceleration through a potential of 20 kV? (a) a proton ($H^+$); (b) a leucine cation ($m = 132$, $z = +1$); (c) a 14.5 kDa protein with $z = +4$. [$e = 1.6 \times 10^{-19}$ C; 1 amu $= 1.66 \times 10^{-27}$ kg]

**3.3.** For each of the above (Problem 3.2), what would be the time-of-flight along a 1.5 metre path?

**3.4.** For each of the 20 kV ions in Problem 3.2, what would be the radius of trajectory in a magnetic field ($B$) of 4 Tesla?

**3.5.** The angular frequency of the orbit of an ion in a magnetic field is given by:

$$\omega = v/r = zeB/m$$

Show how this follows from eqn (3.2).

**3.6.** A recombinant protein (13 700 RMM) has been synthesized as a fusion product with another protein (GST, 12 500 monomer RMM). After cleavage of the fusion protein and (partial) purification, analysis by SDS gel electrophoresis (Chapter 7) shows that, in addition to the expected product with RMM 13 700, there is an additional protein component with RMM of around

26 000–28 000 RMM. Unfortunately, gel electrophoresis is not sufficiently accurate to determine whether this impurity is a dimer of the expected product or uncleaved fusion protein. How might MS be used to resolve this?

## References

1. M. Tyers and M. Mann, From genomics to proteomics, *Nature*, 2003, **422**, 193–197.
2. J. L. P. Benesch and C. V. Robinson, Biological chemistry—dehydrated but unharmed, *Nature*, 2009, **462**, 576–577.

## Further Reading

R. Aebersold and M. Mann, Mass spectrometry-based proteomics, *Nature*, 2003, **422**, 198–207.

E. de Hoffmann, J. Charette and V. Stroobant, *Mass Spectrometry*, Wiley, Chichester, 1996.

A. Dell and H. R. Morris. Glycoprotein structure determination by mass spectrometry, *Science*, 2001, **291**, 2351–2356.

M. Mann, R.C. Hendrickson and A. Pandey, Analysis of proteins and proteomes by mass spectrometry, *Annu. Rev. Biochem.*, 2001, **70**, 437–473.

# 4
# Hydrodynamics

Hydrodynamics is about the movement of water. Both in the bulk liquid and at the molecular level, this can be affected by the presence of other molecules in the liquid. The way in which macromolecules move in solution and the way in which they change the way the solution flows give rise to a range of hydrodynamic techniques that can be used to study their structure and properties.

## Aims

This chapter deals with a variety of hydrodynamic techniques of particular importance in the study of biological macromolecules. By working through this material you should be able to:

- Describe how the density of a liquid mixture may be measured and how it depends on the composition
- Explain the movement of macromolecules during analytical ultracentrifugation and understand the differences between sedimentation equilibrium and sedimentation velocity experiments
- Appreciate the thermal basis for molecular diffusion and Brownian motion
- Describe how viscosity and dynamic light scattering experiments might give information about molecular size

## 4.1    Density and Molecular Volume

The movement of molecules through a fluid depends on the size, shape and mass of the molecules, as well as on the physical properties of the surrounding solvent. Aqueous solutions of proteins, nucleic acids or carbohydrates are generally more dense than pure water under the same conditions. The macromolecules themselves act as if they were denser than water, and this is the basis for the various ultracentrifuge techniques described in this section.

Imagine a mixture made up of molecules ('components') in solution (Figure 4.1). The total volume, $V$, of a mixture of two components

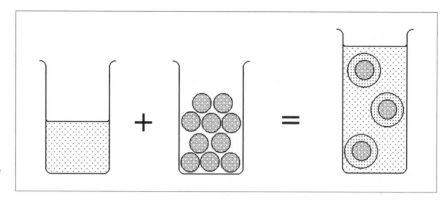

**Figure 4.1** Volumetric effects of mixing.

(where the labels, $1 = \text{solvent}$, $2 = \text{macromolecule}$, say) is given by:

$$V = \bar{V}_1 n_1 + \bar{V}_2 n_2$$

where $n_1$ and $n_2$ are the number of moles of each component in the mixture, and

$$\bar{V}_i = (\partial V / \partial n_i)_{T,P,n_j}$$

is the **partial molal volume** of each component. This can be viewed simply as the increase in total volume (per mole) upon addition of a tiny amount of one component, while keeping everything else (temperature, pressure, other components) constant.

> Partial molal volumes and partial specific volumes are specific examples of a much more general thermodynamic rule: that extensive properties for any system at equilibrium can always be expressed in terms of sums of the partial molal (or partial specific) quantities of each of the individual components in the mixture.

An equivalent expression, using the mass of each component, $g_i$ (in grams), defines the **partial specific volume**, $\bar{v}_i$:

$$V = \bar{v}_1 g_1 + \bar{v}_2 g_2 \quad \text{with} \quad \bar{v}_i = (\partial V / \partial g_i)_{T,P,g_j} = \bar{V}_i / M_i$$

where $M_i$ is the relative molecular mass of component, $i$.

The density, or mass per unit volume, $\rho$, of a mixture is given by:

$$\rho = (g_1 + g_2)/V = (g_1 + g_2)/(\bar{v}_1 g_1 + \bar{v}_2 + g_2)$$

Note that the total volume is not (necessarily) the sum of the volumes of the two separate (pure) components. This can be for several reasons.

> Imagine mixing footballs with ball-bearings . . . The ball-bearings can slide into the spaces between the larger footballs, so the final volume occupied is less than the sum of the volumes when packed separately. [The story is probably apocryphal, but this is said to have been how baseballs (for the troops) and ball-bearings (for the tanks) were shipped efficiently from the US to Europe during World War II.]

First, if the solvent and solute molecules are of different sizes, then there may be voids in the pure components that may become filled when the two are mixed.

Secondly, the solute molecules may affect the structure of the solvent in their vicinity (solvation/hydration effects) and that may affect the overall density of the solution. For example, charged groups or ions in solution will seriously disrupt the normal hydrogen-bonded structure of liquid water, giving a more dense packing of solvent molecules in their hydration shell. Non-polar groups are generally thought to have the opposite effect, as the water molecules tend to

**Table 4.1**  Examples of protein partial specific volumes[1]

| Protein | $M_r$ | $\bar{v}_2/\mathrm{cm}^3\,\mathrm{g}^{-1}$ |
|---|---|---|
| ribonuclease | 13 683 | 0.728 |
| lysozyme | 14 300 | 0.688 |
| ovalbumin | 45 000 | 0.748 |
| serum albumin | 65 000 | 0.734 |
| haemoglobin | 68 000 | 0.749 |
| collagen | 345 000 | 0.695 |
| myosin | 493 000 | 0.728 |

pack into more 'ice-like' clusters in order to accommodate the hydrophobic group.

The observed partial molar volumes of macromolecules (see Table 4.1 for examples) will depend on a number of factors. For globular proteins, for example, it will depend on how well the native protein is folded and how many interior void spaces there might be within the structure, as well as the hydration effects arising from interactions with solvent at the surface.

## Worked Problem 4.1

**Q**: What is the density at 25 °C of a solution made up of 5.0 mg of a protein with a partial specific volume of $0.725\,\mathrm{cm}^3\,\mathrm{g}^{-1}$ in $1.0000\,\mathrm{cm}^3$ of water? The density of pure water at 25 °C ($\rho_0$) is $0.99705\,\mathrm{g\,cm}^{-3}$.

**A**: The mass of $1.0\,\mathrm{cm}^3$ of pure solvent (water), $g_1 = \rho_0 = 0.99705\,\mathrm{g}$.

For the protein: $\bar{v}_2 = 0.725\,\mathrm{cm}^3\,\mathrm{g}^{-1}$ and $g_2 = 0.0050\,\mathrm{g}$.

It is a reasonable approximation for such dilute solutions to assume that the partial specific volume of the solvent (water) in the mixture is the same as for the pure liquid. Consequently, the total volume of the mixture will be given by:

$$V = 1.0000 + \bar{v}_2 g_2 = 1.003625\,\mathrm{cm}^3$$

$$\rho = (g_1 + g_2)/V = (0.99705 + 0.0050)/1.003625$$
$$= 0.99843\,\mathrm{g\,cm}^{-3}$$

Or about 0.14% more dense than pure water.

### 4.1.1  Measurement of Fluid Densities and Partial Volumes

#### 4.1.1.1  Classical Methods

The densities of liquids and liquid mixtures are conventionally measured using **pycnometer** or **hydrometer** methods (Figure 4.2), but these are rarely accurate enough for the dilute macromolecule solutions normally encountered in biophysical studies.

#### 4.1.1.2  Vibrating Tube (Paar) Densimeter

A more precise and convenient method for measuring solution densities, from which partial molal/specific volumes may be calculated, is based on measuring the vibrational frequency of an oscillating quartz tube filled with the solution (Figure 4.3).

The resonant frequency ($\nu$) of any harmonic oscillator is proportional to $(k/M)^{\frac{1}{2}}$, where $M$ is the effective mass of the system and $k$ is a restoring force constant for the oscillator.

The **vibrating tube densimeter** (Paar densimeter) consists of a quartz capillary tube in the form of a loop, clamped at the ends so that it might vibrate like a tuning fork. Filling the tube with fluids of different density causes the resonant frequency of the vibrating tube to change, and this can be measured electronically with great precision. The

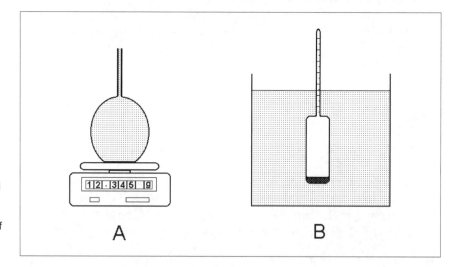

**Figure 4.2**  (A) A pycnometer (or 'specific gravity bottle') is an accurate volumetric flask whose weight change when totally filled with liquid gives the density of that liquid. (B) A hydrometer measures the buoyant density of the fluid by the level at which it floats.

A                    B

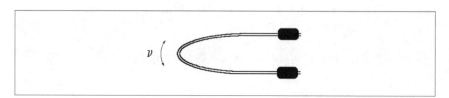

**Figure 4.3**  Oscillating capillary of a vibrating tube densimeter.

period of these oscillations ($\tau = 1/v$) is related to the square-root of the liquid density, or for two fluids of different density:

$$\tau_1^2 - \tau_2^2 = k'(\rho_1 - \rho_2)$$

where $k'$ is a calibration constant that may be determined using standard fluids of known density.

---

**Worked Problem 4.2**

**Q**: Show that $\tau_1^2 - \tau_2^2 = k' (\rho_1 - \rho_2)$

**A**: Assume $v = (k/M)^{\frac{1}{2}}$, so $\tau^2 = 1/v^2 = M/k$

$$M = M_0 + V\rho$$

where $M_0$ is the effective mass of the empty tube and $V$ is its volume. Consequently:

$$\tau^2 = (M_0 + V\rho)/k$$

and for two fluids of different density:

$$\tau_1^2 - \tau_2^2 = (\rho_1 - \rho_2)V/k = k'(\rho_1 - \rho_2)$$

---

### 4.1.1.3 Density Gradient Methods

Archimedes, reputedly while sitting in his bath, showed that the tendency for any object to sink or float is determined by the mass of fluid that the object displaces.

The **buoyant mass** ($m'$) of an object is given by its actual mass ($m_0$) minus the mass of fluid that it displaces:

$$m' = m_0 - m_0 \bar{v}\rho$$

where $\bar{v}$ is the partial specific volume of the object and $\rho$ is the density of the liquid it is displacing.

This is the basis of **density gradient** methods for determining density. Particles sedimenting (falling) through a liquid of increasing density will continue to fall until they reach a region where their buoyant mass is zero and their density matches that of the surrounding solvent. This is usually done in a centrifuge (see below), and the density gradient in the centrifuge tubes can be created using, for example, concentrated solutions of sucrose or caesium chloride (CsCl).

One historically important application of density gradient methods is the classic Meselson and Stahl (1958) experiment[2] in which DNA strands of different densities (incorporating natural $^{14}N$ or enriched with

The Meselson–Stahl experiment confirmed the theory of DNA replication for which Watson and Crick were awarded the Nobel Prize.

[15]N isotopes) were used to demonstrate the semi-conservative replication of the DNA double helix that is central to molecular biology.

## 4.2   Analytical Ultracentrifugation

**Centrifuge methods** for the separation and analysis of biomolecules depend on the centrifugal forces (analogous to gravitational forces but generally much larger) that can be imposed on solutions rotated at high speed. This is the basis of analytical ultracentrifugation (AUC). A centrifuge is simply a device comprising a rotor in which samples may be spun at high speed about a vertical axis. The sample tubes are inserted into compartments in the rotor (either fixed slots or swinging buckets); alternatively, samples can be enclosed in special wedge-shaped cells for analytical methods (Figure 4.4).

Angular velocity, $\omega$, is the rate of rotation in radians per second. One complete rotation (360°) is equal to $2\pi$ radians.

The (outward) force exerted on an object of mass $m$, rotating at a distance $r$ from the axis with angular velocity $\omega$, is equal to $mr\omega^2$. This is equivalent to an acceleration of $r\omega^2$.

---

### Worked Problem 4.3

**Q**: What is the effective acceleration (in $g$) at a radius of 10 cm in a centrifuge rotor spinning at 15 000 rpm?

**A**: 15 000 revolutions per minute (rpm) = $15\,000/60 = 250$ cycles per second, therefore $\omega = 2\pi \times 250 = 1570$ radians per second.

---

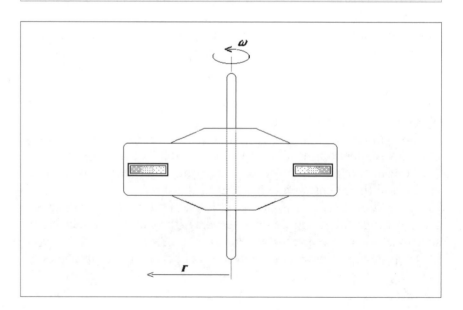

**Figure 4.4** Analytical ultracentrifuge rotor.

$r = 10\,\mathrm{cm} = 0.1\,\mathrm{m}$ (in SI units); $g$ (acceleration due to gravity) $=$ $9.81\,\mathrm{ms}^{-2}$

so the centrifugal acceleration $= r\omega^2 = 0.1 \times 1570^2 = 2.47 \times 10^5$ $\mathrm{ms}^{-2} \equiv 25\,000 \times g$

## 4.3 Sedimentation Equilibrium

Sedimentation equilibrium methods measure the concentration gradients of molecules in solution when spun at high speed in a centrifuge. Samples are held in special cells in the rotor, with optically flat clear windows; in current instruments, several samples (including controls) may be run simultaneously. The concentration gradient, as a function of radius ($r$), is measured using UV/visible absorbance, fluorescence or refractive index, using optical systems mounted outside the rotor.

The tendency for molecules to fall under gravitational or centrifugal forces is offset by thermal diffusion or Brownian motion which tends to disperse them. At equilibrium, the balance between these opposing forces is determined by the Boltzmann probability expression which, for equilibrium sedimentation, means that the relative concentrations of macromolecules at different positions in the centrifuge tube (or cell) can be written as:

$$c(r_1)/c(r_2) = \exp\{-m'(r_1^2 - r_2^2)\omega^2/2RT\} \qquad (4.1)$$

where $m'$ is the **buoyant mass** of the molecule.

## Box 4.1 Sedimentation Equilibrium

Eqn (4.1) arises as follows. The Boltzmann equilibrium expression for the relative probability of finding an object at any point depends on the potential energy difference ($E_1 - E_2$) between the two points, so that $c(1)/c(2) = \exp\{-(E_1 - E_2)/RT\}$.

The force on an object of mass $m$ at radius $r$ is equal to $mr\omega^2$. Consequently, the potential energy difference or work done in moving the object from one radius to another is given by integrating the force $\times$ distance:

$$E_1 - E_2 = \int_{r1}^{r2} mr\omega^2 .dr = m(r_1^2 - r_2^2)\omega^2/2$$

## 4.4    Sedimentation Rate

Unlike sedimentation equilibrium methods, measurement of sedimentation rates (how fast the molecules fall in the centrifugal field) can give information about both macromolecular shape and the size. These methods use relatively high centrifugal forces, and measure the concentration profiles within the centrifuge tube (cell) as a function of time.

Imagine a sedimenting macromolecule. The centrifugal force on the molecule $(m'r\omega^2)$ is opposed by the frictional drag $(F)$ as the molecule tries to move through the surrounding liquid (Figure 4.5). In the case of relatively slow motion (as here), this frictional force is proportional to the velocity $(v)$ of the molecule with respect to solvent:

$$F = fv$$

where $f$ is the 'frictional coefficient' of the molecule. The frictional coefficient $(f)$ depends on a number of factors including the size and shape of the molecule, and the viscosity of the surrounding liquid.

In the steady state, these two forces just balance, so that:

$$fv = m'r\omega^2$$

and the sedimentation rate $v = m'r\omega^2/f$.

In line with expectation, this equation shows that heavier particles (molecules) will sediment faster, while increasing frictional drag $(f)$ will tend to slow them down.

The **sedimentation coefficient** $(s)$, which should be independent of rotor speed or radius, is defined as:

$$s = v/r\omega^2 = m'/f$$

the units of which are seconds. It is more conventional to express sedimentation coefficients in **Svedberg units** (S), in honour of the Swedish scientist T. Svedberg (Nobel Prize 1926) who pioneered much of the early work in this field.

$$1\,S = 10^{-13}\,s$$

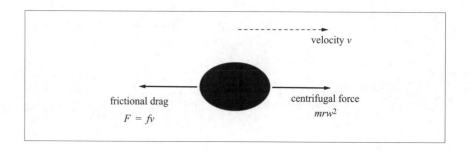

velocity $v$

frictional drag

$F = fv$

centrifugal force

$mrw^2$

**Figure 4.5**  Opposition of frictional and centrifugal forces.

The frictional coefficient depends on the size and shape of the molecules, as well as the temperature and viscosity of the surrounding solvent. This is expressed in the general relationship:

$$f = RT/N_A D$$

where R is the gas constant, $T$ is the absolute temperature, $N_A$ is Avogadro's number, and D is the diffusion coefficient of the macromolecule.

The size and shape dependence of the frictional coefficient is given by the general expression:

$$f = 6\pi\eta R_S$$

where $\eta$ is the viscosity of the solvent, and $R_S$ is the **Stokes radius** of the molecule.

For a truly spherical molecule, the frictional coefficient, $f_0 = 6\pi\eta R_0$, where $R_0$ is the actual radius of the sphere. For other shapes the relationship is more complicated and the deviation from apparent spherical behaviour is expressed by the **frictional ratio**, $f/f_0$. In such cases, the Stokes radius can be viewed as the radius the molecule would have if it were to behave as a sphere. In general, $f/f_0$ is greater than one.[1]

The Stokes radius ($R_S$) will also depend on the solvation of the macromolecule. Even for a perfectly spherical molecule, the solvent molecules in the immediate vicinity will tend to stick to the macro-molecule surface, creating a solvation or hydration shell that will travel with the molecule. Consequently, $R_S$ will often appear larger than the apparent molecular radius determined by other techniques. This can be very useful in estimating the extent of hydration.

A typical sedimentation rate experiment will begin with the sample solution uniformly distributed within the centrifuge cell. However, under the influence of the centrifugal force, any molecules denser than the solvent (water) will begin to move *en masse* towards the bottom of the tube. Typical concentration profiles as a function of time are sketched in Figure 4.6. There are two main features to note. First, the boundary between solvent and solution moves down the tube at a steady rate, and this is the measured sedimentation rate (*v*, measured at the mid-point of the trailing edge).

Secondly, the boundary becomes more diffuse as time goes on. This second feature is due to **diffusion** of the macromolecules at the trailing edge. Although the molecules are all subjected to the same centrifugal forces, each molecule is separately under the influence of random thermal motion from surrounding solvent so that each macromolecule takes a slightly different track, and the edge between solvent and solution becomes more diffuse. This can be analysed to give the diffusion coefficient (*D*) of the macromolecules.

Einstein derived this equation for his 1905 paper on the molecular basis of Brownian motion. During this same year, in an incredible burst of scientific creativity, he also published his famous papers on the quantum theory of the photoelectric effect and on the special theory of relativity.

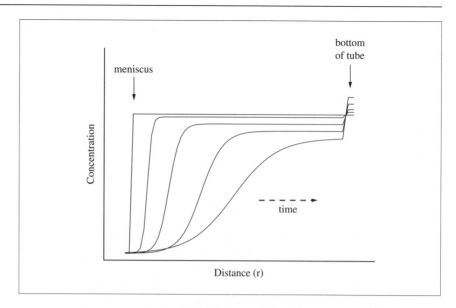

**Figure 4.6** Macromolecular concentration profiles during a sedimentation rate experiment.

Consequently, sedimentation velocity experiments can give both $v$ (and hence $s$) and $D$, which may then be used in the above equations to determine $f$ and $m'$.

### 4.4.1  Mixtures of Molecules

When the sample contains a mixture of molecules with different hydrodynamic properties, the concentration profiles observed in both sedimentation equilibrium and velocity experiments become more complex. However, these can frequently be analysed to determine the individual molecular properties and interactions. For example, a mixture of two different sized macromolecules might, in the simplest case, give two separate boundaries in a sedimentation velocity experiment, each of which might be analysed to give $s$ and $D$ for the separate molecules. If, however, these molecules were to interact strongly (bind to each other) in solution, then this would be seen as a single boundary corresponding to a much larger particle. If the interaction were weaker, involving a dynamic equilibrium between bound and free species, then the boundary shape becomes more diffuse. Techniques are now available to analyse these various models and determine interaction parameters.

### 4.5    Diffusion and Brownian Motion

All molecules are in constant thermal motion and are never completely at rest (except at absolute zero, 0 K). This random molecular motion is what we call 'heat', and objects will feel hot or cold depending on whether their molecular motions are more or less agitated than the surroundings.

We cannot sensibly predict the exact motion of all the molecules in an object, but we can estimate average properties quite exactly. For example, a useful rule of thumb is that the average thermal energy (ignoring quantum effects) is $\frac{1}{2}kT$ per available degree of freedom, so that the average thermal kinetic energy of a gas molecule (3 degrees of freedom) is $3kT/2$ per molecule. [See Chapter 5 for more on this.]

This is a major part of the subject known as 'molecular thermodynamics' or 'statistical mechanics'—see Further Reading.

---

**Worked Problem 4.4**

**Q**: Calculate the average thermal velocity of a water molecule in air at room temperature.

**A**: For a single $H_2O$ molecule (RMM = 18; molar mass = 18 × $10^{-3}$ kg mol$^{-1}$)

$$m = 18 \times 10^{-3}/N_A = 3.0 \times 10^{-26} \text{ kg}$$

$$\tfrac{1}{2}m\langle v^2 \rangle = 3kT/2$$

$$\langle v^2 \rangle = 3kT/m = 3 \times 1.381 \times 10^{-23} \times 300/3.0 \times 10^{-26}$$
$$= 4.1 \times 10^5 \text{ m}^2 \text{ s}^{-2}$$

Root mean square velocity, $\langle v^2 \rangle^{1/2} = 644 \text{ m s}^{-1}$

---

As a consequence of this random thermal motion, any object (large or small) is subject to constant buffeting from its surroundings. This is the source of '**Brownian motion**', the random, chaotic movement of microscopic particles in liquids or gases.

At the macroscopic (everyday) level, we experience this motion as 'atmospheric pressure'. As the above calculations show, the molecules in the air around us are travelling at several hundred metres per second and will be constantly colliding with the surface of our skin. We do not feel each collision (there are too many of them and each impact is, in any case, rather small), but we do feel the sum total of all these collisions as the pressure exerted by the atmosphere.

At the molecular level, this random chaotic movement gives rise to '**diffusion**' and molecules gradually (and irreversibly) get mixed up.

The process of diffusion (in liquids or gases) is described mathematically by **Fick's (first) Law**:

$$\text{Diffusive flux } J = -D.\frac{\partial c}{\partial x}$$

In 1827 the Scottish botanist, Robert Brown, described the perpetual random motion of small particles inside grains of pollen. Although he was not the first to observe this motion, his careful experiments with dead pollen, particles of soot and other inert microscopic objects showed—contrary to popular belief that the grains were alive—that it was a universal property of microscopic particles.

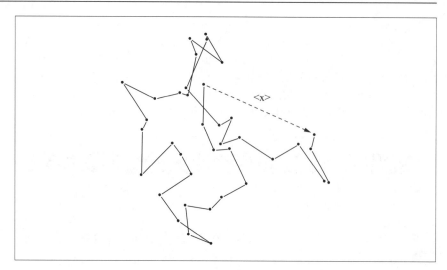

**Figure 4.7**  A diffusional random walk.

This describes how molecules with diffusion coefficient, $D$, will tend to flow in the direction opposite to any concentration gradient $\frac{\partial c}{\partial x}$.

At the molecular level this may be viewed 8s a **random walk** in which the molecules move in a series of random, uncoordinated steps (Figure 4.7). In 1905, Einstein used this concept to derive the equation that describes how particles will move under such motion. The mean square displacement (distance moved) of a spherical particle in any direction over time, $t$, is given by:

$$\langle x^2 \rangle = 6Dt$$

where $\langle x^2 \rangle$ is the mean square displacement in any direction.

This equation accurately describes the Brownian motion of microscopic particles and was used by Jean-Baptiste Perrin in a classic series of experiments (from 1908) to estimate Avogadro's number ($N_A$). He was awarded the Nobel Prize in physics in 1926.

The diffusion coefficient ($D$) for a macromolecule depends on its size, shape and flexibility, and the viscosity and temperature of the surrounding fluid. Consequently, it is generally difficult to estimate theoretically, but a number of models have been developed. In general the diffusion coefficient is just another way of expressing the frictional coefficient ($f$):

$$D = RT/N_A f = kT/f$$

so that, for an idealized spherical particle with (Stokes) radius, $R_S$:

$$D = RT/6\pi N_A \eta R_S \, (\text{Stokes} - \text{Einstein equation})$$

**Worked Problem 4.5**

**Q**: The diffusion coefficient measured in water at 20 °C for the protein lysozyme is about $10.4 \times 10^{-11} \, m^2 \, sec^{-1}$. What is $R_S$ for this molecule given that the viscosity of water at 20 °C is $1.002 \times 10^{-3} \, N \, s \, m^{-2}$?

**A**: Rearranging the above equation:

$$R_S = RT/6\pi N_A \eta D$$
$$= 8.314 \times 293/(6\pi \times 6 \times 10^{23} \times 1.002 \times 10^{-3} \times 10.4 \times 10^{-11})$$
$$= 2.1 \times 10^{-9} \, m \, (2.1 \, nm)$$

Note that, in addition to translational motion, other kinds of motion will be subject to random thermal fluctuations. In particular, the tumbling of (macro)molecules in solution can be described in terms of **rotational diffusion**, for which equivalent rotational diffusion constants can be defined. By analogy with the translational diffusion described above, the rotational diffusion coefficient ($D_{rot}$) and rotational friction coefficient ($f_{rot}$) are related by the expression:

$$D_{rot} = kT/f_{rot}$$

For rotation of an ideal sphere of radius, $R$, in a fluid of viscosity, $\eta$, it has been shown that:

$$f_{rot} = 8\pi\eta R^3$$

It is often useful to imagine how fast a macromolecule might be rotating or tumbling in solution. Obviously such motion will be quite chaotic, and only average values can be estimated, but this is given by the **rotational relaxation time**:

$$\tau_{rot} = 1/2D_{rot}$$

Which, for an ideal sphere, becomes: $\tau_{rot} = 8\pi\eta R^3/2kT$.

**Worked Problem 4.6**

**Q**: What is the rotational relaxation (tumbling) time for a globular protein of radius 1 nm in water at 20 °C? (The viscosity of pure water at 20 °C is $1.002 \times 10^{-3} \, N \, s \, m^{-2}$.)

$$\mathbf{A:}\ \tau_{rot} = 8\pi\eta R^3/2kT$$
$$= 8\pi \times 1.002 \times 10^{-3} \times (1 \times 10^{-9})^3/(2 \times 1.381 \times 10^{-23} \times 293)$$
$$= 3 \times 10^{-9}\ s(3\,ns)$$

## 4.6    Dynamic Light Scattering (DLS)

The diffusional or Brownian motion of molecules in a liquid or gas gives rise to fluctuations in density or concentration that can be observed by optical methods.

Imagine a small volume element in a solution of macromolecules. At any moment in time, some molecules will be diffusing into this volume while others will be diffusing out (Figure 4.8). If the volume is large enough or the observation period is long enough, then these numbers moving in or out will cancel each other out. However, for small volumes over short time intervals, this will not always be the case and fluctuations in density or concentration will occur. For a beam of light passing through the sample, these fluctuations will be seen as fluctuations in refractive index and some of the light will be scattered. This scattered light will appear to 'twinkle' as the number of molecules in the volume element fluctuates up or down. The rate at which this twinkling occurs will depend (amongst other things) on the rate at which the molecules are diffusing in solution.

This is the basis of the technique known as '**dynamic light scattering**' (DLS). A laser beam is passed through the solution of macromolecules and the time dependence of the light scattered from a small volume within the sample is recorded (Figure 4.9).

Analysis of the shape and frequency of this flickering pattern gives the '**autocorrelation time**' ($\tau$) which is related to the diffusion constant (D) of the molecules. This information can be used to determine the relative molecular masses and heterogeneities of macromolecular samples.

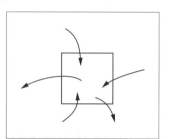

**Figure 4.8**   Diffusion in and out of a small volume element.

Density fluctuations in the Earth's atmosphere are responsible for the familiar blue colour of the (cloudless) sky. White light from the Sun hitting the upper atmosphere is scattered (in all directions) by these fluctuations in volume elements about the size of the wavelength of light. Since the relative magnitude of the density fluctuations is bigger for smaller volume elements, shorter wavelengths (blue light) tends to get scattered more than longer wavelengths (red light). What we see when looking up into the sky is this scattered blue part of the Sun's spectrum. Conversely, looking directly into the Sun at sunrise or sunset (do not do this at any other times!) we see the redder end of the spectrum after the blue light has been scattered out of the beam. Pollution (small particles in the atmosphere) tends to enhance this effect.

## 4.7    Viscosity

Solutions of macromolecules tend to be more viscous than the pure solvent, and this forms the basis for one of the earliest experimental methods for determining the shape properties of biomolecules. There are a number of ways to measure solution viscosities, but the simplest uses a **capillary (or Ostwald) viscometer** (Figure 4.10). Other techniques include the **falling sphere** method, which involves measuring how fast a heavy sphere falls through the liquid, and the **Couette viscometer**

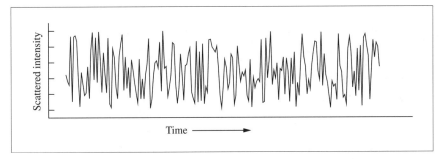

**Figure 4.9** Example of fluctuating intensity of scattered light in a DLS experiment.

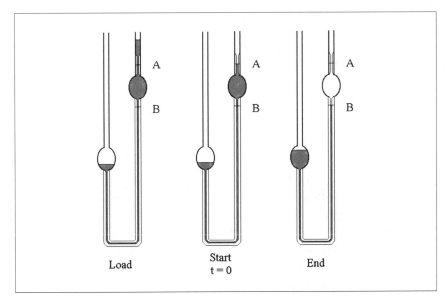

Load

Start
t = 0

End

**Figure 4.10** Capillary viscometry.

which measures the torque or force generated between concentric rotating cylinders with the test solution filling the gap.

The rate of flow under gravity of a liquid through a capillary tube depends on a number of factors including the viscosity ($\eta$) and density ($\rho$) of the liquid as well as the size and shape of the tube. For a standard capillary viscometer (see Figure 4.10), the time taken ($t$) for a set volume of liquid to flow between points A and B is proportional to $\eta/\rho$ so that, after appropriate calibration with known liquids, the viscosity of any sample can be determined from its flow time, $t$.

Viscosities are still sometimes given in non-SI units called 'poise, P', named after the French physician, Jean Poiseuille (1797–1869), who developed a method for measuring blood pressure and who was responsible for fundamental studies of liquid flow. $1\,P = 0.1\,N\,s\,m^{-2}$.

---

### Worked Problem 4.7

**Q**: The flow time ($t_0$) for water at 20 °C in a capillary viscometer was 27.3 seconds. For a dilute protein solution under the same conditions, $t = 30.4$ sec. What is the viscosity of the protein solution?

> **A**: The viscosity of pure water at $20\,^{\circ}C$ is $\eta_0 = 1.002 \times 10^{-3}\,N\,s\,m^{-2}$.
>
> Assuming that the liquids have (almost) identical densities (reasonable for dilute biomolecular solutions):
>
> $$t/t_0 = \eta/\eta_0$$
>
> so $\eta = \eta_0 t/t_0 = 1.002 \times 10^{-3} \times 30.4/27.3 = 1.118 \times 10^{-3}\,N\,s\,m^{-2}$.

*Viscosity and its effect on the flow of liquids through very narrow tubes ('capillaries') is important in the design of miniaturized microfluidics devices, ink-jet printer heads and 'Lab-on-Chip' assays.*

Viscosities are often expressed relative to the viscosity of the pure solvent by introducing the terms **relative viscosity** ($\eta_r$), **specific viscosity** ($\eta_{sp}$) and **intrinsic viscosity** ($[\eta]$), defined as follows:

$$\text{Relative viscosity}: \quad \eta_r = \eta/\eta_0$$
$$\text{Specific viscosity}: \quad \eta_{sp} = (\eta - \eta_0)/\eta_0 = \eta_r - 1$$
$$\text{Intrinsic viscosity}: \quad [\eta] = \eta_{sp}/c \ (\text{as } c \to 0)$$

*Viscosity is commonly used to determine whether samples of DNA in solution are single- or double-stranded, and to detect conformational changes in double-helical DNA when binding potential drug molecules.[3]*

where $c$ is the concentration of macromolecules. Because of molecular interactions and other non-ideality effects in solution, $[\eta]$ is normally determined by extrapolation to zero concentration from a range of measurements (Figure 4.11).

The relationship between $[\eta]$ and macromolecular shape is difficult to determine theoretically, except in special cases, but a number of empirical rules have been derived.[1]

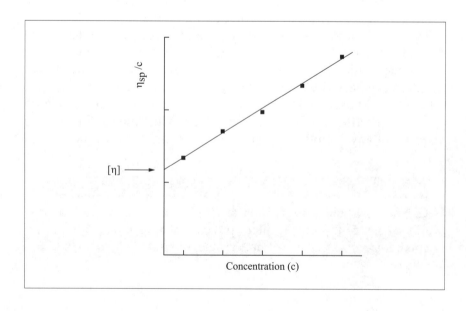

**Figure 4.11**  Variation of intrinsic viscosity, $\eta_{sp}/c$, with concentration.

## Summary of Key Points

1. Aqueous solutions of macromolecules are more dense than pure water. Their density is determined by the partial molar volumes of the different components.
2. The sedimentation of macromolecules in high speed centrifugation can be used to estimate the shape, size and homogeneity of macromolecules in solution.
3. Translational and rotational diffusion of molecules in solution, and the related Brownian motion of larger particles, is due to random thermal motions and can be related to size and shape.
4. Additional hydrodynamic properties such as viscosity may also give size and shape information.

## Problems

**4.1.** 7.5 mg of a protein were dissolved in 5.000 g water and the density of the resulting solution at 25 °C was found to be $0.99748\,\mathrm{g\,cm^{-3}}$. What is the partial specific volume of this protein under these conditions? [The density of pure water at 25 °C, $\rho_1 = 0.99707\,\mathrm{g\,cm^{-3}}$.]

**4.2.** This protein unfolds irreversibly when heated. After such treatment, the density of the solution when cooled down to 25 °C changes to $0.99752\,\mathrm{g\,cm^{-3}}$. What does this indicate about the changes in partial specific volume upon unfolding?

**4.3.** Why might the partial specific volume of a protein decrease upon unfolding?

**4.4.** When running samples in a centrifuge, why is it necessary to ensure that the weight is distributed symmetrically around the rotor?

**4.5.** High speed centrifuges are heavily armour plated. Why?

**4.6.** Calculate the rotational kinetic energy ($\approx \frac{1}{2}mr^2\omega^2$) of a 2 kg centrifuge rotor, with effective radius of 15 cm, spinning at 40 000 rpm. The explosive energy of TNT is about $4.6 \times 10^6\,\mathrm{J\,kg^{-1}}$. Compare.

**4.7.** Estimate the diffusion coefficients ($D$) in water at 20 °C for the following:

(a) a spherical molecule with radius 0.3 nm;

(b) a globular protein of radius 2.5 nm;

(c) a bacterial cell of diameter 10 μm.

(The viscosity of water, $\eta = 1.002 \times 10^{-3} \, \mathrm{N \, s \, m^{-2}}$ at 20 °C.)

**4.8.** Under the same conditions, how far (on average) would the objects in problem 4.7 travel in water under Brownian motion over a five-minute interval?

**4.9.** Why are the above distances (Problem 4.8) smaller than might be expected from the thermal velocities of these particles under the same conditions?

## References

1. *C. Tanford, Physical Chemistry of Macromolecules*. New York:Wiley, 1961.
2. M. Meselson and F. W. Stahl, The replication of DNA in *Escherichia coli*, *Proc. Natl. Acad. Sci. U. S. A.*, 1958, **44**, 671–682.
3. K. M. Guthrie, A. D. C. Parenty, L. V. Smith, L. Cronin and A. Cooper, Microcalorimetry of interaction of dihydro-imidazo-phenanthridinium (DIP)-based compounds with duplex DNA, *Biophys. Chem.*, 2007, **126**, 117–123.

## Further Reading

J. M. Seddon and J. D. Gale, *Thermodynamics and Statistical Mechanics*, RSC Tutorial Chemistry Text, Royal Society of Chemistry, Cambridge, 2001.

D. Sheehan, *Physical Biochemistry: Principles and Applications*, Wiley, New York, 2nd edn, 2009, ch. 7.

I. Tinoco, K. Sauer, J. C. Wang and J. D. Puglisi, *Physical Chemistry: Principles and Applications in Biological Sciences*, Prentice Hall, Upper Saddle River, NJ, 4th edn, 2002, ch. 6.

K. E. van Holde, W. C. Johnson and P. S. Ho, *Principles of Physical Biochemistry*, Prentice Hall, Upper Saddle River, NJ, 1998, ch. 5.

# 5

# Thermodynamics and Interactions

Like all material things, the structure and behaviour of biological systems are governed by the interplay between the thermal motion of molecules and the various interaction forces between them. This chapter reviews the basics of thermodynamic equilibrium at the molecular level, leading to ways in which we can measure the various components of thermodynamic forces in biomolecular systems.

## Aims

After working through this chapter you should be able to:

- Explain the molecular basis of heat and thermodynamics
- Describe microcalorimetric methods for the direct determination of the thermodynamics of biomolecules
- Show how spectroscopic and other methods may be used indirectly to estimate thermodynamic parameters
- Explain equilibrium dialysis methods for measuring binding equilibrium
- Describe the thermodynamic basis for protein solubility and crystallization

## 5.1  A Bluffer's Guide to Molecular Thermodynamics

Thermodynamics is a mature subject and it is sometimes difficult not to get bogged down in the details. The basic ideas were laid down in the early 19th century, at a time when atoms and molecules had scarcely been heard of and when optimization of the steam engine and other marvels of the industrial revolution were of prime concern. What follows here, though less rigorous than you will find in most textbooks, will be enough to place thermodynamics in a molecular context.

Thermodynamic equilibrium represents the balance between the natural tendency for things to fall to lower energy (Figure 5.1), together

'**Work**' is the organized motion of atoms and molecules. '**Heat**' is the disorganized motion of atoms and molecules.

The internal energy ($U$) of a system is just the total kinetic and potential energy of all the atoms and molecules in the sample including translational motions, rotations and vibrations, and the forces between them.

with the equally natural tendency for random thermal motion at the molecular level (*i.e.* heat) to oppose this (Figure 5.2).

Most things in the everyday world take place at constant pressure, so the appropriate energy quantity to use is **enthalpy**, $H = U + PV$, which comprises the **internal energy** ($U$) plus a pressure–volume corrections term ($PV$) to take account of any work done with the surrounding if there is a change in volume. Changes in enthalpy during any process will be indicated by $\Delta H$.

The disrupting effects of thermal motion at the molecular level are expressed in terms of **entropy** changes, $\Delta S$. These effects will tend to get bigger as the temperature rises.

The balance between these two opposing effects is expressed in the change in **Gibbs free energy**:

$$\Delta G = \Delta H - T.\Delta S$$

At thermodynamic equilibrium when these two effects just balance, $\Delta G = 0$ and, although the individual atoms and molecules are still in motion, the overall average situation undergoes no further spontaneous change.

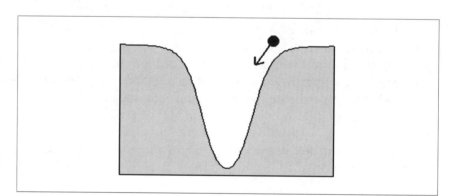

**Figure 5.1**  Things tend to want to roll downhill . . . $\Delta H$ tends to be negative.

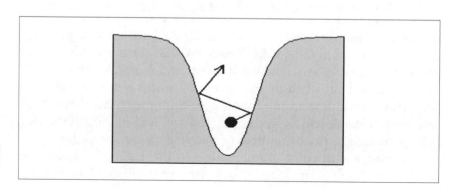

**Figure 5.2**  Thermal (Brownian) motion tends to kick things uphill . . . $\Delta S$ tends to be positive.

In **statistical thermodynamics,** recognizing that we can only talk about the probabilities of things happening at the molecular level, these effects are expressed in the **Boltzmann probability:**

$$p(H) = w . \exp(-H/RT)$$

where $p(H)$ is the probability that a system of molecules will have an enthalpy $H$ (expressed per mole) at a temperature T (in K), and R is the universal gas constant (R $= 8.314\,\mathrm{J\,K^{-1}\,mol^{-1}}$). The quantity, $w$, sometimes known as the **degeneracy,** is the number of different ways in which the system can adopt the enthalpy, $H$. As we shall see shortly, it is related to entropy.

The exponentially decreasing term in the Boltzmann expression would seem to favour the very lowest energy states. However, this would lead to the paradoxical situation in which everything in the Universe should be at zero enthalpy. This can be resolved as follows . . .

The total energy of the Universe is not zero. At the instant of the Big Bang (or whatever), the world was endowed with a large amount of energy. This energy will never go away, no matter how hard we try. So it must be distributed somehow. But how does that fit with the Boltzmann expression? It's all to do with '$w$'.

There are not many ways you can get low energies. At very low energies things don't move around very much, molecules tend to crystallize and everything is nice and ordered; '$w$' is small.

But at higher energies, things are much more exciting. Molecules move around a lot, or rotate or vibrate—lots of things happening at once. So there are lots of different ways of getting the same energies; '$w$' is big.

The Boltzmann constant is related to the gas constant and Avogadro's number: $k = R/N_A = 1.381 \times 10^{-23}\,\mathrm{J\,K^{-1}}$.

The combination of these two effects can be seen graphically (Figure 5.3).

The average thermal energy of molecular motion is related to the absolute temperature, $T$. For example, the mean thermal kinetic energy of any object (molecule) of mass, $m$, is given by:

$$\tfrac{1}{2} m \langle \mathrm{v}^2 \rangle = 3kT/2$$

where $\langle \mathrm{v}^2 \rangle$ is the mean square velocity of the molecule and k is the Boltzmann constant.

The situation is a little more complicated for molecular rotations and vibrations, where quantum effects need to be taken into account. But, for low energy motions, the general rule is that the average thermal energy is about $\tfrac{1}{2} kT$ per degree of freedom.

## Worked Problem 5.1

**Q**: What is the average thermal velocity of a molecule of nitrogen ($N_2$) in the air around us?

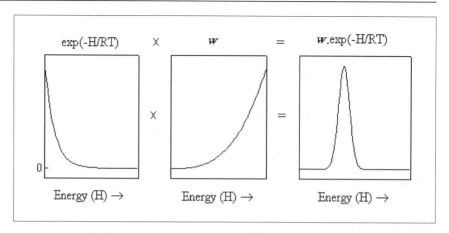

**Figure 5.3** The product of '$w$' (increasing with energy) × the exponential (decreasing with energy) gives a probability distribution that means that most objects spend most of their time at some average energy away from zero.

**A:** For a single $N_2$ molecule (RMM $= 28$; molar mass $= 28 \times 10^{-3}$ kg mol$^{-1}$):

$$m = 28 \times 10^{-3}/N_A = 4.7 \times 10^{-26} \text{ kg}$$

$$\langle v^2 \rangle = 3kT/m = 3 \times 1.381 \times 10^{-23} \times 300/4.7 \times 10^{-26}$$
$$= 2.6 \times 10^5 \text{ m}^2 \text{ s}^{-2}$$

Root mean square velocity, $\langle v^2 \rangle^{1/2} = 514$ m s$^{-1}$, *i.e.* about half a kilometre per second.

You might be wondering: if the molecules in the air around us are moving so fast, how come we don't feel the effects of them colliding with us? Well, we do. We call it atmospheric pressure.

Such thermal motion is not, of course, all in a straight line —or at least for not very long, since such rapidly moving molecules soon collide with their neighbours and the surroundings, so that the motion is a much more chaotic random walk. This is why molecules diffuse. It is also why proteins and other biological macromolecules are quite dynamic structures, constantly undergoing fluctuations due to collisions from their surroundings.

### 5.1.1　Chemical Equilibrium

Consider a simple chemical equilibrium $A \rightleftharpoons B$

The reaction never really stops (at the molecular level), but equilibrium is reached when the rate of the forward reaction just matches that of the reverse reaction. Thus we have a dynamic equilibrium in which the situation can be described in terms of the **probabilities** of the molecules being in one state or the other (A or B) at any one time.

What is the relative probability of finding molecules in either state A or state B?

Using the Boltzmann rule:

$$\text{Probability of state A, } p(A) = w_A . \exp(-H_A/RT)$$
$$\text{Probability of state B, } p(B) = w_B . \exp(-H_B/RT)$$

where $w_A$ and $w_B$ are the 'numbers of ways' for each molecular form with enthalpies $H_A$ and $H_B$, respectively.

So combining these two equations, the ratio of the relative probabilities of the two chemical forms is:

$$p(B)/p(A) = \exp(-\Delta H^\circ/RT) \times w_B/w_A, \text{ where } \Delta H^\circ = H_B - H_A$$

However, this ratio is what we would normally call the **equilibrium constant**, $K$, for this reaction:

$$K = [B]/[A] = p(B)/p(A) = \exp(-\Delta H^\circ/RT) \times w_B/w_A$$

Now let's do some algebraic rearrangement to get this into something more familiar. First, take natural logarithms (ln) of both sides of this equation:

$$\ln K = -\Delta H^\circ/RT + \ln(w_B/w_A)$$

Remember the general rules for logarithms and exponentials:

$$\ln[\exp(y)] = y$$
$$\ln(a \times b) = \ln a + \ln b$$

Now multiply both sides by $-RT$ to give:

$$-RT . \ln K = \Delta H^\circ - RT . \ln(w_B/w_A)$$

Providing we make the connection, $\Delta S^\circ = R.\ln(w_B/w_A)$, this result is identical to the classical definition of the **standard** Gibbs free energy change:

$$\Delta G^\circ = \Delta H^\circ - T.\Delta S^\circ = -RT . \ln K$$

This shows how entropy is actually related to the degeneracy, $w$, or the number of ways in which molecular systems can exist with a particular energy.

## 5.1.2 Heat Capacity

Both enthalpy and entropy are fundamentally related to the **heat capacity** (or **specific heat**) of an object.

$$\Delta H(T) = \Delta H(T_{ref}) + \int_{Tref}^{T} \Delta C_p . dT$$

$$\Delta S(T) = \Delta S(T_{ref}) + \int_{Tref}^{T} (\Delta C_p/T) . dT$$

where $\Delta C_p$ is the heat capacity change at constant pressure, which is related to the temperature dependence of both entropy and enthalpy:

$$\Delta C_p = \partial \Delta H / \partial T = T.\partial \Delta S / \partial T$$

Heat capacity is the quantity which tells us how much heat energy ($H$) we need to add to a system in order to increase its temperature. It is also related to entropy ($S$) because, if $w$ is high, there are lots of different ways in which the added energy might be distributed without raising the temperature and the heat capacity is consequently also high.

Both enthalpy ($\Delta H$) and entropy ($\Delta S$) changes are important in determining chemical and physical equilibrium. This is particularly the case in biomolecular systems dominated by non-covalent interactions, where the intermolecular forces are relatively small and comparable to entropic effects arising from molecular dynamics and solvation changes.

The consequence of all this is that, if we want to understand how biological molecules behave, we must somehow determine the different thermodynamic contributions to the forces that control their structure and interactions. This is primarily an experimental problem and the following sections show how it can be done in various ways.

## 5.2   Differential Scanning Calorimetry

Differential scanning calorimetry (DSC) is an experimental technique to measure directly the heat energy uptake that takes place in a sample during controlled increase (or decrease) in temperature. At the simplest level it may be used to determine thermal transition ('melting') temperatures for samples in solution, solid, or mixed phases (*e.g.* suspensions). But with more sensitive apparatus and more careful experimentation, it may be used to determine absolute thermodynamic data for thermally induced transitions of various kinds. It is particularly useful in studying the thermodynamics of unfolding transitions in dilute solutions of proteins and nucleic acids, or the phase transitions that can take place in biological membranes.

Figure 5.4 shows the typical layout of a DSC instrument. Sample and reference solutions are contained in identical calorimetric vessels, or cells (labelled S and R), typically around 1 cm$^3$ in volume. In a DSC experiment, the sample solution, S (typically a protein at a concentration of 1 mg cm$^{-3}$ or less in modern instruments) is heated at constant rate in the calorimeter cell alongside an identical reference cell (R) containing buffer. Both sample and reference solutions are kept under a small positive pressure, $P$ (1–2 atm), to inhibit bubble

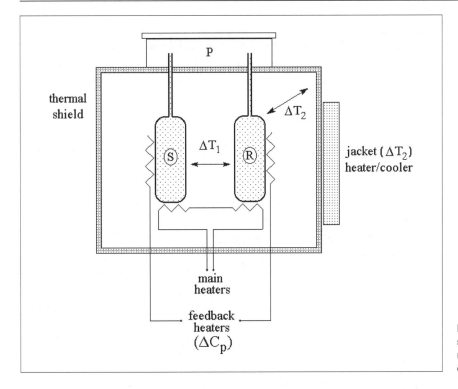

**Figure 5.4** A differential scanning calorimeter for measuring thermal transitions in dilute solution.

formation from dissolved gases as the temperature is increased. The temperature differences between S and R ($\Delta T_1$) and the surrounding jacket ($\Delta T_2$) are measured by sensitive thermocouples, which provide a voltage proportional to the temperature difference that may be sensed and amplified by external electronics. The entire system is heated at constant rate by the jacket and main heaters, and each cell (R and S) can be heated separately using feedback heaters. The power supplied to these heaters (voltage and current) is measured and recorded.

Picture what happens as the sample and reference solutions are heated up at a constant rate. Initially, if both sample and reference behave the same, there will be no temperature difference between them. But, at some temperature, the protein molecules in the sample solution (for example) may begin to thermally unfold and some of the heat energy from the main heaters will be used to bring about this endothermic transition rather than in raising the temperature. Consequently, there will be a temperature lag ($\Delta T_1$) between the sample and reference cells. This is detected by the external electronics and additional heat is supplied to S (using the feedback heater) to correct this imbalance. The electrical heat energy supplied to the sample, in this case, is a direct measure of the enthalpy change in the sample due to the temperature change.

---

**Worked Problem 5.2**

**Q:** What might be the temperature difference between sample and identical buffer reference solutions for a sample comprising $1 \, \text{mg} \, \text{cm}^{-3}$ of a protein of RMM 50 000 undergoing a thermal transition with $\Delta H = 80 \, \text{kJ} \, \text{mol}^{-1}$?

**A:** $1 \, \text{mg} \equiv 1 \times 10^{-3}/50 \, 000 = 2 \times 10^{-8}$ moles of protein $\times \Delta H$ $\equiv 1.6 \times 10^{-3} \, \text{J}$ heat energy uptake per mg of protein.

The specific heat capacity of water (assume identical for buffer and protein solution) $= 4.2 \, \text{J} \, \text{K}^{-1} \, \text{cm}^{-3}$

Assuming that all this heat energy is absorbed by the $1 \, \text{cm}^3$ sample, $\Delta T_1 = 1.6 \times 10^{-3}/4.2 = 3.8 \times 10^{-4} \, °\text{C}$

(In practice, thermal transitions in biomolecules do not occur all at once but take place over a finite temperature range. This means that temperature changes observed by DSC are usually very much smaller than this.)

---

In general, the differences in heat energy uptake between the sample and reference cells required to maintain equal temperature correspond to differences in apparent heat capacity. It is these differences in heat capacity that give direct information about the energetics of thermally induced processes in the sample.

An example of the sort of data obtained by DSC for the thermal unfolding of a small protein in water is shown in Figure 5.5. This shows the measured heat capacity of a protein in solution as a function of temperature. At low temperatures the heat capacity ($C_p$) is relatively small. However, above about $40 \, °\text{C}$ in this case, the protein begins to unfold endothermically. This requires heat energy and the heat capacity curve rises sharply, reaching a maximum at the mid-point temperature of the transition ($T_m$) before falling to a new baseline level once all the protein molecules are unfolded.

The integrated area under the $C_p$ transition peak gives the total heat energy (enthalpy) required to bring about the observed transition. But the shape of the curve gives experimental information as well. For example, the fraction of protein unfolded at any temperature can be estimated from the relative areas under the curve (Figure 5.6).

The increase in baseline heat capacity ($\Delta C_p$) is characteristic. It shows that the heat capacity of the unfolded polypeptide chain is greater than that of the folded protein. This is typical of what one sees during the melting of hydrogen bonded solids (Figure 5.7), and is also

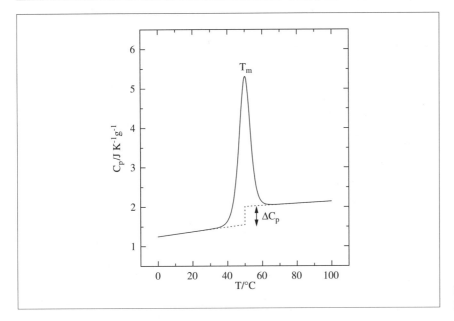

**Figure 5.5** Typical DSC data—protein unfolding.

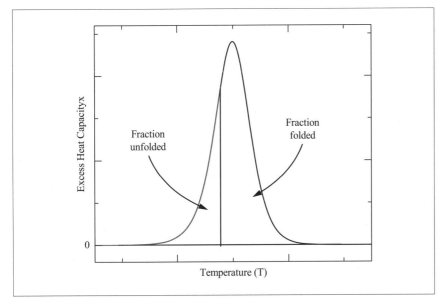

**Figure 5.6** Integrated areas under a DCS peak are related to extent of the transition.

what is expected for the increased exposure to water of hydrophobic amino acid side chains in the unfolded state.[1]

## 5.3   Isothermal Titration Calorimetry

The thermodynamics of interaction between molecules in solution can be measured using isothermal titration calorimetry (ITC) (Figure 5.8)

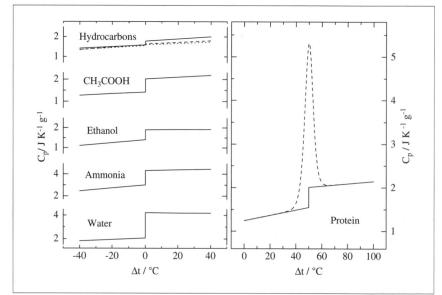

**Figure 5.7** (Left) Examples of absolute heat capacities for pure solid and liquid compounds as a function of temperature, plotted relative to the normal melting point ($\Delta t = T - T_m$). (Right) Protein unfolding data (in aqueous solution) for comparison plotted on a similar heat capacity scale.

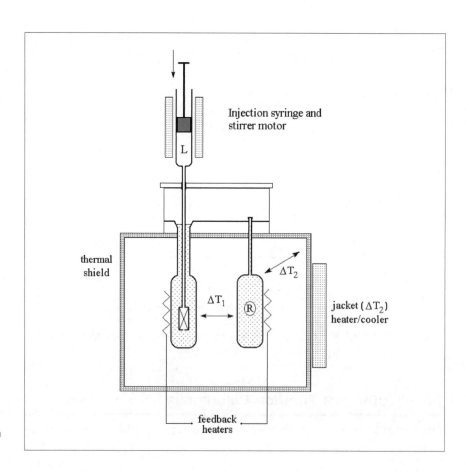

**Figure 5.8** Isothermal titration calorimeter.

The ITC instrument is very similar to the DSC described above. In this case, however, instead of changing the temperature, the sample and reference are kept at constant temperature (hence 'isothermal'). The sample cell is fitted with an injection syringe so that small amounts of another solution can be mixed with the sample. Typically this method might be used to measure the heat of binding of a drug or inhibitor molecule to an enzyme.

Typical ITC data for the binding of a small inhibitor molecule to an enzyme are shown in Figure 5.9. Note how initially quite large heat pulses are observed with each injection. This corresponds to the heat energy liberated when the inhibitor molecules bind to the protein active site. With subsequent injections, however, the heat pulses get smaller as eventually all the available binding sites are occupied. General methods for analysing data such as these are described in Section 5.5.

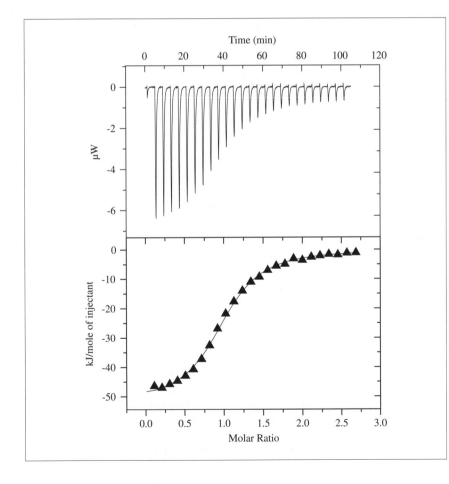

**Figure 5.9** Typical ITC data for binding of a trisaccharide inhibitor (tri-$N$-acetyl-glucosamine; tri-NAG) to hen egg white lysozyme, in $0.1\,mol\,dm^{-3}$ ethanoate buffer, pH 5. Each exothermic heat pulse (upper panel) corresponds to injection of $10\,\mu l$ $(0.01\,cm^3)$ of tri-NAG $(0.45\,mmol\,dm^{-3})$ into the protein solution $(36\,\mu mol\,dm^{-3})$. Integrated heat data (lower panel) constitute a differential binding curve that may be fit to a standard single-site binding model to give, in this instance,[2] the stoichiometry of binding (number of binding sites), $N = 0.99$; binding affinity, $K_{ass} = 3.9 \times 10^5\,(mol\,dm^{-3})^{-1}$ $(K_{diss} = 2.6\,\mu mol\,dm^{-3})$, and enthalpy of binding, $\Delta H = -51.7\,kJ\,mol^{-1}$.

## 5.4    Binding Equilibrium

Suppose we know the equilibrium constant for binding of a ligand to a protein. How do we know how much is bound under particular conditions? Typically we might know the total protein and total ligand concentrations, but how much is bound?

For protein–ligand binding (or anything equivalent):

$$P + L \rightleftharpoons PL$$

The dissociation constant is:

$$K = [P][L]/[PL] \tag{5.1}$$

and the total ligand concentration is:

$$C_L = [L] + [PL] \tag{5.2}$$

Now the total protein concentration:

$$C_P = [P] + [PL]$$

Using equation (5.1) this becomes:

$$C_P = K[PL]/[L] + [PL]$$

And using equation (5.2):

$$C_P = K[PL]/(C_L - [PL]) + [PL]$$

Rearrange to give the quadratic equation for [PL]:

$$[PL]^2 - (C_P + C_L + K)[PL] + C_P C_L = 0$$

For which the two solutions are:

$$[PL] = [(C_P + C_L + K) \pm \{(C_P + C_L + K)^2 - 4C_P C_L\}^{1/2}]/2$$

By inspection, the minus sign is the physically correct solution, giving the exact expression for protein–ligand complex formation, [PL], as a function of the total protein and ligand concentrations:

$$[PL] = [(C_P + C_L + K) - \{(C_P + C_L + K)^2 - 4C_P C_L\}^{1/2}]/2$$

If there are $n$ binding sites per mole of protein, then $C_P = n.C_0$, where $C_0$ is the estimated protein concentration, giving:

$$[PL] = [(n.C_0 + C_L + K) - \{(n.C_0 + C_L + K)^2 - 4nC_0 C_L\}^{1/2}]/2$$

The fraction, $\phi$, of sites occupied at any ligand concentration is given by:

$$\phi(C_L) = [PL]/n.C_0$$
$$= [(n.C_0 + C_L + K) - \{(n.C_0 + C_L + K)^2$$
$$- 4nC_0C_L\}^{1/2}]/2nC_0 \qquad (5.3)$$

Binding to multiple binding sites on biological macromolecules often involves more complex binding schemes, possibly involving **cooperativity** or **allosteric interactions** between binding sites. These situations give rise to more complicated binding expressions that are not explored here.

## 5.5    General Methods for Determining Thermodynamic Properties

There are many different ways in which the changes in conformation of a macromolecule or binding of ligands can be observed experimentally. Some of these methods, such as the calorimetric techniques described above, give thermodynamic information directly. Other methods, many of them based on spectroscopic changes (see Chapter 2), are more indirect, but we can still obtain useful thermodynamic data provided we have a reasonable idea about what is going on in the process.

For example, increasing temperature or adding chemical 'denaturants' such as urea or guanidinium hydrochloride (GuHCl) produce changes in fluorescence or the circular dichroism (CD) spectra of proteins in solution consistent with unfolding of the protein structure (Figures 5.10 and 5.11).

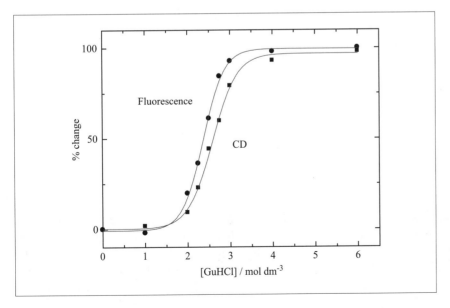

**Figure 5.10** Effects of a chemical denaturant (GuHCl) on the intrinsic fluorescence and circular dichroism intensity of a protein in solution.

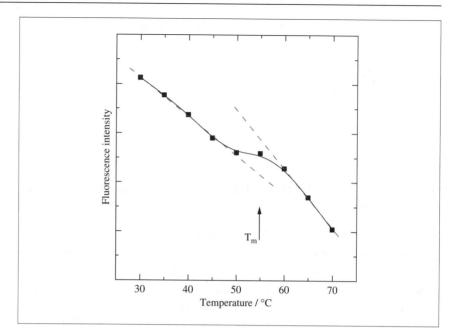

**Figure 5.11** Fluorescence emission intensity *versus* temperature for a typical protein (lysozyme) in solution. The inflection point ($T_m$) indicates thermal unfolding at about 55 °C in this example. The dashed lines indicate the general decrease in Trp fluorescence emission with increase in temperature for both the folded (low T) and unfolded (high T) states.

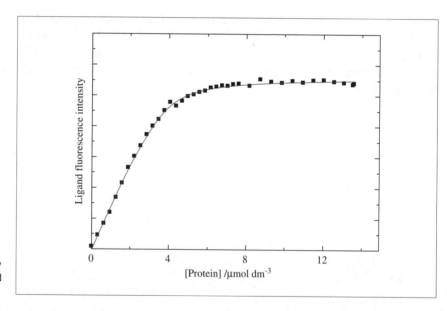

**Figure 5.12** Binding of a fluorescence-labelled fatty acid (dansyl undecanoic acid) to a lipid-binding protein. Addition of increasing amounts of protein causes an increase in fluorescence emission of the ligand. The fluorescence intensity reaches a plateau when all ligand molecules are bound.

Binding of ligands to DNA strands or protein active sites can also result in spectral changes (either in the ligand or the macromolecule) that depend on the extent of binding. See Figure 5.12 for an example.

How might we analyse such data to obtain useful thermodynamic information? The following sections illustrate how this can be done.

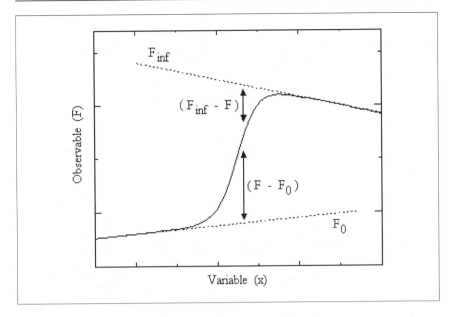

**Figure 5.13** General two-state transition curve. The observable (F) can be any experimental parameter (heat uptake, fluorescence intensity, UV absorbance, CD intensity, *etc.*). The variable (*x*) is whatever is bringing about the transition (temperature, pressure, concentration, pH, *etc.*). The dotted lines give the extrapolated baselines, showing what F would be for the initial or final states in the absence of any transition.

### 5.5.1 Folding/Unfolding Transitions

For a two-state unfolding transition:

$$N \rightleftharpoons U, \; K = [U]/[N]$$

The equilibrium constant at any particular temperature can be determined from the experimental data in Figure 5.13 as follows:

$$K = (F - F_0)/(F_{inf} - F)$$

From which the free energy for unfolding can be calculated in the standard way:

$$\Delta G^{\circ}_{unf} = \Delta H^{\circ}_{unf} - T.\Delta S^{\circ}_{unf} = -RT.\ln K$$

Measurements of $K$ across the temperature range can be used to derive $\Delta H^{\circ}_{unf}$ and $\Delta S^{\circ}_{unf}$.

---

**Worked Problem 5.3**

**Q**: In the example in Figure 5.14, $F_0 = 50$ and $F_{inf} = 75$. What are the $K$ and $\Delta G^{\circ}_{unf}$ at 35 °C?

**A**: At 35 °C, F (from the graph) = 53.8.

$$K(35°C) = (F - F_0)/(F_{inf} - F) = (53.8 - 50)/(75 - 53.8)$$
$$= 3.8/21.2 = 0.179$$

$$\Delta G^\circ_{\text{unf}} = -RT. \ln K = -8.314 \times (273 + 35) \times \ln(0.179)$$
$$= +4.4\,\text{kJ}\,\text{mol}^{-1}$$

The same approach may be used to determine $K$ and $\Delta G^\circ_{\text{unf}}$ at other temperatures giving the following results (check these for yourself):

| $T/^\circ\text{C}$ | F | $K$ | $\Delta G^\circ/\text{kJ}\,\text{mol}^{-1}$ |
|---|---|---|---|
| 35 | 53.8 | 0.18 | +4.4 |
| 40 | 62.5 | 1 | 0 |
| 45 | 71.9 | 7.1 | −5.2 |

Note here that, at the mid-point of the transition, $T_{\text{m}} = 40\,^\circ\text{C}$, $K = 1$ and $\Delta G^\circ_{\text{unf}} = 0$.

A rough estimate of the enthalpy and entropy of unfolding can be made using $\Delta G^\circ_{\text{unf}} = \Delta H^\circ_{\text{unf}} - T.\Delta S^\circ_{\text{unf}}$ with data at two different temperatures. For example:

At 35°C :  $\Delta G^\circ_{\text{unf}} = \Delta H^\circ_{\text{unf}} - 308.\Delta S^\circ_{\text{unf}} = +4.4\,\text{kJ}\,\text{mol}^{-1}$

At 45°C :  $\Delta G^\circ_{\text{unf}} = \Delta H^\circ_{\text{unf}} - 318.\Delta S^\circ_{\text{unf}} = -5.2\,\text{kJ}\,\text{mol}^{-1}$

Using the method of simultaneous equations, subtracting one from the other gives:

$$10 \times \Delta S^\circ_{\text{unf}} = +9.6\,\text{kJ}\,\text{mol}^{-1}$$

Hence:  $\Delta S^\circ_{\text{unf}} = +0.96\,\text{kJ}\,\text{mol}^{-1}\,\text{K}^{-1}$

We can now use this value to estimate $\Delta H^\circ_{\text{unf}}$ by substitution into one of the free energy equations. For example:

At 35°C :  $\Delta G^\circ_{\text{unf}} = \Delta H^\circ_{\text{unf}} - 308 \times 0.96 = +4.4\,\text{kJ}\,\text{mol}^{-1}$

Hence:  $\Delta H^\circ_{\text{unf}} = +300\,\text{kJ}\,\text{mol}^{-1}$

(Bear in mind that these are only approximate estimates since $\Delta H$ and $\Delta S$ often vary with temperature.)

### 5.5.2  Ligand Binding

For simple ligand binding equilibrium (such as described in Section 5.4), the extent of binding is proportional to the change in experimental observable. Consequently, for any particular total ligand

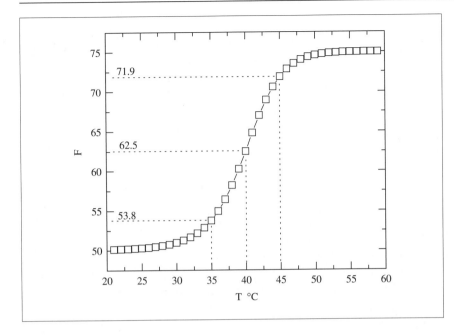

**Figure 5.14** The graph shows some typical data for the UV absorbance (in arbitrary units) of a macromolecule that undergoes an unfolding transition with increase in temperature. This is the sort of curve one might see for the 'melting' of a DNA double helix, for example.

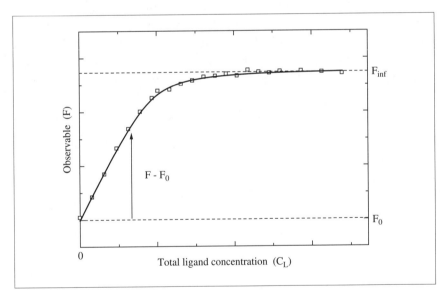

**Figure 5.15** Ligand binding titration curve. This shows the change in some experimental observable (F) with increasing ligand concentration. Eventually F reaches a constant plateau value ($F_{inf}$) when all binding sites are full.

concentration, $C_L$, we can determine the fraction bound from the experimental curve (Figure 5.15).

The fraction, $\phi$, of sites occupied at any ligand concentration is given by:

$$\phi(C_L) = [PL]/C_P = (F - F_0)/(F_{inf} - F_0)$$

Unfortunately, since we don't know the free ligand concentration ([L]) only the total ($C_L$), it is necessary to use a fairly complicated expression—such as eqn (5.3) from Section 5.4—to estimate K from this. However, in cases of relatively weak binding, when [PL] is small compared to the total ligand concentration, then we might make the approximation that [L] $\approx$ $C_L$.

## 5.6    Thermal Shift Assays

It is sometimes necessary to screen large numbers of compounds for potential binding to specific biomolecules, but the techniques described so far are often too slow or require too much material for this to be achieved economically on a realistic timescale. In addition, in the first stages of a drug discovery project, for example, it is not always necessary to have a full thermodynamic characterization—simply asking the question 'does it bind or not?' is enough.

A thermal shift assay is based on the principle that binding of a ligand to the native state of a protein (say) will increase the thermal stability of the protein (see Box 5.1). This arises because additional energy is required to displace the ligand from the protein before it can unfold. The resulting increase in $T_m$ may be observed by various techniques, and an example using DSC is given in Figure 5.16.

Other techniques more suited to rapid screening of multiple samples have been developed, in particular taking advantage of large changes

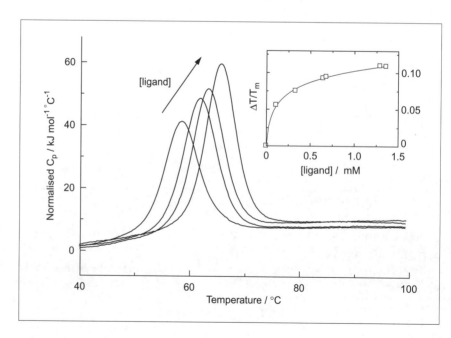

**Figure 5.16**  Example of DSC thermograms for protein unfolding in the presence of various concentrations of ligand, illustrating the increase in $T_m$ brought about by binding of ligand to the native protein. The inset shows the relative $T_m$ shift as a function of total ligand concentration (adapted from ref. 3).

in the fluorescence of probe molecules that bind to unfolded proteins (Section 2.4.5). Thermal unfolding of the protein in the presence of such probes gives a large increase (usually) in fluorescence intensity, and this can be used to estimate $T_m$ for the protein. This is simple to detect and has been adapted for use with small samples in multi-well formats, allowing simultaneous screening of large numbers of compounds for potential binding.[4]

It is sometimes assumed (incorrectly) that the increase in thermal stability of a protein–ligand complex is due to a conformational change in the native protein. No. It is simply a consequence of a shift in folded ⇌ unfolded equilibrium.

---

## Box 5.1 Ligand Binding and Protein Folding Equilibrium

For a simple case in which a ligand molecule (L) binds specifically to a single site on the native folded protein (N), the following equilibria apply:

$$\text{Ligand binding}: \quad N + L \rightleftharpoons NL, \quad K_{L,N} = [N][L]/[NL]$$

$$\text{Unfolding}: \quad N \rightleftharpoons U, K_0 = [U]/[N]$$

where $K_{L,N}$ is the dissociation constant for ligand binding to the native protein and $K_0$ is the unfolding equilibrium constant for the protein in the absence of ligand.

The effective protein unfolding equilibrium constant ($K_{unf}$) is given by the ratio of the total concentrations of unfolded to folded species:

$$K_{unf} = [U]/([N] + [NL]) = K_0/(1 + [L]/K_{L,N}) \approx K_0 K_{L,N}/[L]$$

where the final approximation applies only at high free ligand concentrations ($[L] > K_{L,N}$). This confirms the expectation that $K_{unf}$ decreases and the folded form becomes more stable with increasing ligand concentration, since this will shift thermodynamic equilibrium in favour of the folded form.

The shift in $T_m$ can be estimated from:

$$\Delta T_m/T_m = (RT_{m0}/\Delta H_{unf,0}) \times \ln(1 + [L]/K_L)$$

in which $\Delta T_m = T_m - T_{m0}$ is the change in unfolding transition temperature and $\Delta H_{unf,0}$ is the enthalpy of unfolding of the protein in the absence of bound ligand.[3]

Note that similar effects will be observed if the ligand binds instead to the unfolded form of the protein, though in this case the $T_m$ will be reduced.

Pigs' bladders or other biological membranes were often used for this purpose. Nowadays, **semi-permeable** or **dialysis** membranes are made of synthetic materials in which the pore size can be more carefully controlled. The same principle is used in kidney dialysis and other therapeutic applications to remove small molecule toxins while leaving cells and macromolecules intact.

## 5.7    Equilibrium Dialysis

One problem with the indirect methods for ligand binding described in the previous section is that we rarely know the actual concentrations of free ligand or complex, for example. Rather, these are inferred from indirect observations. Equilibrium dialysis is a technique that gets around this problem in some instances.

A simple equilibrium dialysis cell is made up of two compartments in which the solutions are separated by a semi-permeable membrane (Figure 5.17). This membrane contains small channels that allow free diffusion of solvent and small solute molecules, but through which larger molecules such as proteins cannot pass. In a typical experiment the protein (or other macromolecule) solution is placed in one compartment with solvent (buffer) in the other, and ligand is added. The (small) ligand molecules (L) are free to move through the membrane, but the protein (P) and the protein ligand complex (PL) are confined to one side. Once equilibrium is reached, the experiment involves measurement of the *total* protein and *total* ligand concentrations in each of the compartments.

With the configuration given in Figure 5.16:
For the right-hand compartment:

(a)  Total concentration of L: $C_L(\text{right}) = [L]$
     For the left-hand compartment:
(b)  Total concentration of L: $C_L(\text{left}) = [L] + [PL]$
(c)  Total concentration of P: $C_P(\text{left}) = [P] + [PL]$

Hence $[PL] = $ (b) $-$ (a) and $[P] = $ (c) $- [PL]$ and $K = [PL]/[P][L]$.

Equilibrium dialysis methods often involve the use of radioactively labelled ligands, so that very small concentrations and very tight binding can be measured directly.

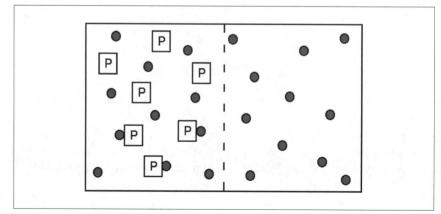

**Figure 5.17**  Equilibrium dialysis cell, showing the macromolecules (P) and complex confined to one side, whilst the smaller ligand molecules are free to diffuse across the semi-permeable membrane (indicated by the dashed line).

## 5.8    Protein Solubility and Crystallization

In addition to the fairly specific interactions we have considered so far, there are numerous non-specific interactions that make proteins and polypeptides quite 'sticky', with a tendency to adhere both to each other and to other surfaces. As a consequence, proteins and polypeptides usually have quite low solubilities that are sensitive to changes in pH, ionic strength and other solutes. This is important in both experimental and practical applications, since most proteins are only properly active in solution. Some general guidelines . . .

A saturated solution is what we get when equilibrium is reached between the solid phase of a compound and its solution in a particular solvent. As always, this is a battle between the tendency for thermal motion to shake apart any aggregates, opposed by any net attractive forces holding the aggregates together. At low enough concentrations, the thermal disruptive (entropic) effects always win (given time), because dilute solutions or mixtures have a high entropy. At high concentrations, however, a limit is reached wherein no further solute will dissolve. In thermodynamics, this can be related to the standard free energy of transfer from the solid to solution ($\Delta G^\circ_{solution}$):

$$\Delta G^\circ_{solution} = -\mathrm{R}T \cdot \ln(\text{solubility})$$

The **solubility** of any component is the concentration of solute ($\mathrm{mol\,dm}^{-3}$) in the solution phase at equilibrium and in contact with the solid phase.

Anything that increases this free energy will reduce the solubility and *vice versa*.

### 5.8.1  Electrostatic Effects on Solubility

Electrostatic interactions between charged groups on proteins appear to be one of the main factors controlling their solubility in aqueous systems.

The charges on proteins are due (mainly) to the weakly acidic and basic side chains, and consequently depend on the pH of the solution (see Chapter 1). Proteins tend to carry either an overall positive charge (low pH) or negative charge (high pH), and the resulting electrostatic repulsions will tend to overcome any more specific attractive interactions between the molecules. However, at intermediate pH values, particularly at the isoelectric point when pH = pI, the proteins carry a net zero charge and this overall repulsive effect will be minimized (see Figure 5.18). Hence, as a general (but not universal) rule, proteins tend to be least soluble at their isoelectric point.

### 5.8.2  'Salting-in' and 'Salting-out'

The effects of added electrolytes (salts) on protein solubility are complex. At low concentrations (typically with salt concentrations

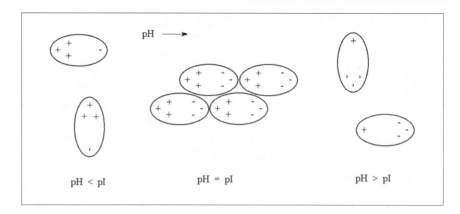

pH ⟶

pH < pI          pH = pI          pH > pI

**Figure 5.18** Effect of pH on electrostatic attractions/ repulsions between proteins.

below about $0.5 \, \text{mol} \, \text{dm}^{-3}$), electrostatic screening by the small ions in solution tends to reduce electrostatic interactions between the macromolecules, and the solubility of protein *increases*. Historically this is known as the 'salting-in' effect. However, at higher salt concentrations the solubility of protein tends to *decrease* due to the 'salting-out' effect.

Salting-out is an indirect thermodynamic effect due more to the strong affinity of electrolytes for water rather than to specific interactions between proteins. Small, highly charged ions are strongly solvated in polar solvents such as water. They prefer a high dielectric environment and will tend to repel non-polar molecules.

There are two (roughly equivalent) ways of looking at this from a thermodynamic viewpoint. Dissociation of salts into small ions in solution occurs more readily in a polar environment—salts 'prefer' to be in a very polar medium like water. The presence of non-polar material such as protein will tend to reduce the overall polarity of the medium, and will be thermodynamically unfavourable for small ions. Consequently, especially at high salt concentrations, non-polar substances such as proteins will be forced out of solution so as to maximize the overall polarity of the medium. Alternatively, one may imagine that at high salt concentrations there is insufficient water available to solvate both the small ions and the protein surface. Consequently, protein is forced out of solution to release more water molecules for solvation of the salts.

Both salting-in and salting-out effects depend on the size and charge of the ions in solution. Particularly for salting out, small highly charged ions tend to have a greater effect (because of their higher solvation). This has given rise to the **Hofmeister** or **lyotropic series** comparing the relative effectiveness of different anions or cations on salting out. A typical Hofmeister series, in decreasing order of

Although the surfaces of protein molecules tend to be relatively polar, the main bulk of the protein is made up of non-polar material of low dielectric constant.

effectiveness in salting out proteins from aqueous solution, is given below for common anions and cations.

Anions:  $citrate^{2-} > SO_4^{2-} > HPO_4^{2-} > F^- > Cl^- > Br^- > I^- > NO_3^- > ClO_4^-$

Cations:  $Al^{3+} > Mg^{2+} > Ca^{2+} > Na^+ > K^+ > Cs^+ > NH_4^+ > N(CH_3)_4^+$

(Note that the actual sequence can vary slightly depending on protein and other conditions.)

Salting-out or precipitation by varying salt type and concentration is often used in the early stages of purification of proteins from complex mixtures.

This empirical series arises from experiments performed in 1888 by the Austro-German chemist Franz Hofmeister (1850–1922) who was interested in the relative effectiveness of different salts in precipitating proteins from egg white. The alternative name, lyotropic series, derives from the Greek meaning roughly 'pertaining to solvent affinity'.

### 5.8.3   Non-polar Additives

Since solubility in water is generally mediated by the polar groups on the protein surface, it follows that any reduction in solvent polarity might tend to reduce protein solubility. This is generally the case. Addition of non-polar solutes such as ethanol, methanol or polymers such as polyethylene glycol (PEG) usually lowers the protein solubility (Figure 5.19).

Some additives, especially at high concentrations, may also unfold or otherwise denature the protein, and this may lead to protein precipitation for other reasons.

### 5.8.4   Protein Crystallization

Proteins normally precipitate out of solution as amorphous aggregates. However, under carefully controlled conditions, single crystals of purified proteins may be obtained. This is a crucial step in the determination of protein structure by diffraction methods (see Chapter 8).

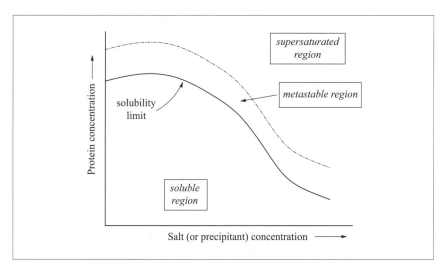

**Figure 5.19**  Phase diagram illustrating the effects of salt or other precipitant concentration on protein solubility.

'**Nucleation**' is the initial
formation of a small aggregate
from which the larger crystal may
grow. Since crystal nuclei involve
just a few molecules with
relatively few intermolecular
contacts, they tend to be
relatively unstable and do not
form spontaneously.

The process can be described by reference to a phase diagram (Figure 5.19). At low concentrations of protein and salt (or other precipitant), the protein is below the solubility limit and remains in solution. At high concentrations, however, the solution becomes thermodynamically unstable (supersaturated) and, once nucleation takes place, protein will precipitate out of solution until the concentration in the liquid phase falls back to the solubility limit. If this happens too quickly, the precipitate tends to be amorphous or just composed of very tiny crystals.

Nucleation is the key to more controlled crystal growth. In the supersaturated region of the phase diagram, nucleation is easy but haphazard, and subsequent crystal growth is rapid and hard to control. However, in the 'metastable' region, at concentrations slightly above the saturation limit, small nuclei are unstable but larger nuclei may grow at the expense of smaller ones. Most protein crystallization methods attempt to achieve this by slowly increasing the protein and precipitant concentrations until the solution approaches this metastable region. Several practical methods have been devised, mostly based on liquid or vapour phase diffusion processes. The 'sitting drop' technique is one example (Figure 5.20).

In the sitting drop method, a small volume of protein solution is mixed with an equal volume of precipitant solution (which may be a concentrated solution of salts or other precipitants) and placed in a chamber surrounded by a larger volume of the precipitant solution. Water will evaporate from the protein droplet and pass through the vapour phase into the more concentrated precipitant solution. This gradually increases the concentrations of both the protein and the precipitants in the drop until the precipitant concentration equals that of the surrounding reservoir. Under the right conditions (which are usually achieved by trial and error), this slow increase in concentration (typically over many days) allows growth of well formed crystals.

Screening for optimal
crystallization conditions is time-
consuming and labour-intensive.
Much of this is now done
routinely using automated robotic
systems.

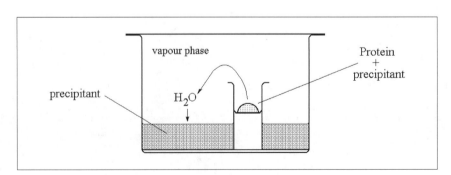

**Figure 5.20** Sitting drop method for protein crystallization.

## Summary of Key Points

1. The conformation and interactions of biological macro-molecules are governed by thermodynamic forces involving a balance of entropic and enthalpic effects.
2. The thermodynamics of biomolecular stability and inter-actions can be measured directly using microcalorimetry methods.
3. Spectroscopic and other indirect methods can also be used to determine thermodynamic properties.
4. Equilibrium dialysis methods can be used to measure binding interactions directly.
5. The solubility of proteins (and other macromolecules) can be understood in terms of thermodynamic phase equilibrium.

## Problems

**5.1.** What is the average thermal velocity of the following:

(a) an oxygen molecule at 25 °C;

(b) a water molecule at 25 °C;

(c) a protein molecule of RMM 25 000 at 37 °C?

**5.2.** How might we feel or otherwise detect such motion in the molecules around us?

**5.3.** Thermal stability studies have given the following (partial) thermodynamic data for unfolding of a protein in aqueous so-lution at pH 7.4 at different temperatures:

(a) Complete this table by supplying the missing data (?) where possible.

(b) What fraction of the protein molecules would be un-folded at 50, 55 and 60 °C, respectively, under these conditions?

(c) What does the temperature dependence of the unfolding enthalpy ($\Delta H^\circ$) suggest about the forces responsible for stabilizing the folded protein conformation?

| $t/^\circ C$ | $K$ | $\Delta G^\circ/\text{kJ mol}^{-1}$ | $\Delta H^\circ/\text{KJ}^{-1}\text{mol}^{-1}$ | $\Delta S^\circ/\text{K}^{-1}\text{mol}^{-1}$ |
|---|---|---|---|---|
| 45 | 0.133 | 5.33 | 150.0 | ? |
| 50 | ? | 2.86 | 175.0 | ? |
| 55 | ? | 0 | 200.0 | 609.8 |
| 60 | 3.22 | ? | 225.0 | ? |

**5.4.** The complete genome sequence of a simple nematode worm (*Caenorhabditis elegans*) has been completed. One of the major tasks now is to identify the function of many of the gene products. Scientists have identified one protein that might have metal-binding properties. Describe different biophysical techniques that might be used to investigate the binding of metal ions to this protein in solution.

**5.5.** The following experimental data have been obtained for the fluorescence intensity (F) and circular dichroism intensity (CD) of a protein solution at different temperatures.

(a) What is the $T_m$ of this protein?

(b) What fraction of the protein might be unfolded at 46 °C?

(c) What is the Gibbs free energy of unfolding at this temperature?

(d) Explain what molecular properties are being monitored by the two different sets of data.

(e) Do the transitions monitored by fluorescence and CD necessarily have to occur at the same temperature? If not, explain why not.

| $T/^\circ C$ | F (arbitrary units) | CD (arbitrary units) |
|---|---|---|
| 20 | 65.0 | − 1310 |
| 30 | 65.0 | − 1310 |
| 40 | 64.7 | − 1304 |
| 46 | 58.8 | − 1186 |

| 50 | 40.0 | $-810$ |
| 56 | 17.8 | $-366$ |
| 60 | 15.5 | $-320$ |
| 70 | 15.0 | $-310$ |
| 80 | 15.0 | $-310$ |

**5.6.** For binding of a ligand (L) to a protein (P) to form a 1 : 1 complex (PL), show that: $c_p/[PL] = 1 + 1/K[L]$, where $K$ is the equilibrium constant and $c_p$ is the total protein concentration. Explain how this expression might be used to analyse equilibrium dialysis data.

**5.7.** In an equilibrium dialysis experiment to study the binding of a new organic ligand to a soluble receptor protein, the following data were obtained:

Left-hand (protein + ligand) compartment:

Total protein concentration   $= 8.3 \times 10^{-9} M$

Total ligand concentration   $= 3.9 \times 10^{-8} M$

Right-hand (ligand only) compartment:

Total ligand concentration   $= 3.5 \times 10^{-8} M$

What is the equilibrium binding constant for binding of the ligand?

### References

1. A. Cooper, Heat capacity of hydrogen-bonded networks: an alternative view of protein folding thermodynamics, *Biophys. Chem.*, 2000, **85**, 25–39.
2. A. Cooper, C. M. Johnson, J. H. Lakey and M. Nollmann, Heat does not come in different colours: entropy-enthalpy compensation, free energy windows, quantum confinement, pressure perturbation calorimetry, solvation and the multiple causes of heat capacity effects in biomolecular interactions, *Biophys. Chem.*, 2001, **93**, 215–230.
3. A. Cooper, M. A. Nutley and A. Wadood, Differential scanning microcalorimetry, in *Protein–Ligand Interactions: Hydrodynamics and Calorimetry*,

ed. S. E. Harding and B. Z. Chowdhry, Oxford University Press, Oxford, 2000, pp. 287–318.

4. M. W. Pantoliano, E. C. Petrella, J. D. Kwasnoski, V. S. Lobanov, J. Myslik, E. Graf, T. Carver, E. Asel, B. A. Springer, P. Lane and F. R. Salemme, High-density miniaturized thermal shift assays as a general strategy for drug discovery, *J. Biomol. Screen.*, 2001, **6**, 429–440.

## Further Reading

Cooper, Microcalorimetry of protein-protein interactions, in *Biocalorimetry: Applications of Calorimetry in the Biological Sciences*, ed. J. E. Ladbury and B. Z. Chowdhry, Wiley, Chichester, 1998, pp. 103–111.

Cooper, Microcalorimetry of protein-DNA interactions, in *DNA-Protein Interactions*, ed. A. Travers and M. Buckle, Oxford University Press, Oxford, 2000, pp. 125–139.

Cooper, Heat capacity effects in protein folding and ligand binding: a re-evaluation of the role of water in biomolecular thermodynamics, *Biophys. Chem.*, 2005, **115**, 89–97.

G. A. Holdgate, Making cool drugs hot: isothermal titration calorimetry as a tool to study binding energetics, *Biotechniques*, 2001, **31**, 164–186.

D. Sheehan, *Physical Biochemistry: Principles and Applications*, Wiley, New York, 2nd edn, 2009, ch. 8.

J. M. Seddon and J. D. Gale, *Thermodynamics and Statistical Mechanics*, RSC Tutorial Chemistry Text, Royal Society of Chemistry, Cambridge, 2001.

K. E. van Holde, W. C. Johnson and P. S. Ho, *Principles of Physical Biochemistry*, Prentice Hall, Upper Saddle River, NJ, 1998, ch. 2–4, 15.

I. Tinoco, K. Sauer, J. C. Wang and J. D. Puglisi, *Physical Chemistry: Principles and Applications in Biological Sciences*, Prentice Hall, Upper Saddle River, NJ, 4th edn, 2002, ch. 2–4.

# 6
# Kinetics

Thermodynamics describes what should happen at equilibrium. Kinetics tells us how fast we are getting there. Living things are never at thermodynamic equilibrium: biology succeeds because of the very careful way in which the rates of biochemical processes are controlled.

## Aims

In this chapter we look at some of the experimental methods used to measure the rates of biomolecular reactions, together with a revision of background theory. By the end, together with previous studies, you should be able to:

- Explain the factors that affect how fast a reaction proceeds
- Describe basic methods for measuring reaction rates
- Describe methods for following fast reactions
- Understand the basic kinetic equations of enzyme catalysis and inhibition

## 6.1    Basic Kinetics

Any chemical or physical system not already at thermodynamic equilibrium will tend to move to restore that equilibrium. But the manner in which it does this, and how fast it gets there, depends on numerous factors. Rates may be extremely fast like the explosive reaction of hydrogen with oxygen, or imperceptibly slow like the erosion of marble by acid rain. Reaction kinetics are dealt with more rigorously elsewhere. Here we just summarize the basics in order to set the scene for later sections.

In order for any reaction to proceed at the molecular level, in general three conditions must be fulfilled: (a) molecules must collide, in (b) a correct orientation, with (c) sufficient energy to overcome the activation barrier to reaction (Figure 6.1).

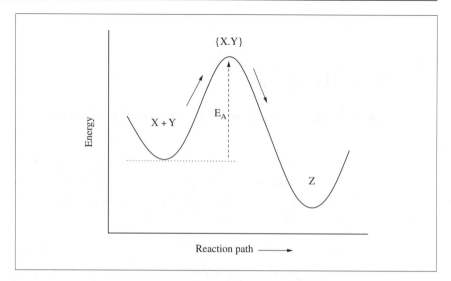

**Figure 6.1** Typical energy landscape for a simple chemical reaction pathway. Reactant molecules (X and Y) must come together with sufficient energy to surmount the activation barrier ($E_A$) and form an activated complex, or intermediate, {X.Y}, before proceeding to product (Z).

These conditions are expressed in the basic rate equations. Consider, for example, the elementary reaction:

$$X + Y \rightarrow Z$$

The **rate of reaction** at any time is defined as the overall rate of formation of products or loss of reactants. For homogeneous reactions in solution, this is expressed in terms of the concentrations of reactants or products:

$$\text{Rate} = d[Z]/dt = -d[X]/dt = -d[Y]/dt$$

The **rate equation** for this elementary reaction would be:

$$\text{Rate} = k[X][Y]$$

where k is the **rate constant** and [X] and [Y] are the molar concentrations of reactants. It is important for these basic equations to express the concentrations in molar units (moles per $dm^3$ rather than, for example, grams per $dm^3$) because this expresses the way in which changing the actual number of molecules in the mixture will directly affect the probability of intermolecular collisions.

The rate constant, k, encapsulates the other factors required for reaction, expressed in the classic **Arrhenius equation**:

$$k = A . \exp(-E_A/RT) \tag{6.1}$$

where the pre-exponential quantity, $A$, is a **collision factor** that depends on frequency and orientation of the colliding molecules, and $E_A$ is the **activation energy** for the reaction. The exponential energy term expresses the (Boltzmann) probability that the colliding molecules will have sufficient energy to overcome the activation energy barrier at any given absolute temperature (T).

The temperature dependence of the rate constant is sometimes expressed in terms of a more rigorous **transition-state theory** which, although strictly correct only for ideal gas reactions, can be useful in some circumstances. The relevant equation is:

$$k = (k_B T / h) \times \exp(-\Delta G^{\#}/RT)$$
$$= (k_B T / h) \times \exp(-\Delta S^{\#}/R) \times \exp(-\Delta H^{\#}/RT) \qquad (6.2)$$

where $\Delta G^{\#} = \Delta H^{\#} - T.\Delta S^{\#}$ is the **activation free energy**, made up of the **activation enthalpy** and **activation entropy**, and can be pictured as the thermodynamic free energy required to form the transition state complex for the reaction. The activation enthalpy, $\Delta H^{\#}$ is analogous to the Arrhenius activation energy, $E_A$, whereas the orientation and other factors appear explicitly in the pre-exponential as the activation entropy, $\Delta S^{\#}$.

$k_B$ is the Boltzmann constant, h is the Planck constant and R is the gas constant.

Reactions can go backwards as well, and the overall reaction rate is a balance between the forward and reverse reactions. For the elementary reaction:

$$X + Y \underset{k_2}{\overset{k_1}{\rightleftharpoons}} Z$$

$$\text{Overall rate} = k_1[X][Y] - k_2[Z]$$

Equations (6.1) and (6.2) don't work at very low temperatures approaching absolute zero (0 K). This arises because of quantum mechanical barrier penetration (or 'tunnelling'), where particles can pass through barriers without requiring thermal energy to climb over them. This is thought to be significant in some enzymic reactions, in particular those involving electron or hydrogen transfer.[1]

Equilibrium is reached when the rates of forward and backward reactions just balance, and the overall rate is zero. Consequently, when the concentrations have reached their equilibrium values:

$$k_1[X]_{equilib}[Y]_{equilib} - k_2[Z]_{equilib} = 0$$

which rearranges to give:

$$k_1/k_2 = [Z]_{equilib}/[X]_{equilib}[Y]_{equilib} \equiv K$$

where K is the **equilibrium constant** for the reaction.

More complex reaction mechanisms may proceed in a sequence of elementary steps that cannot be predicted, necessarily, from the overall balanced equation. This may involve **transient reaction intermediates** and **rate-limiting steps** that control the observed kinetics. Consequently, a more general rate expression is of the form:

Reactions don't stop when equilibrium is reached. The individual molecules continue to react in both forward and backward directions, but the overall concentrations no longer change.

See Chapter 5 for more on equilibrium. Remember that K will usually have units, unless the powers of the concentration terms in the numerator and denominator exactly cancel.

$$\text{Rate} = k[X]^m[Y]^n$$

where m and n are the **order of reaction** with respect to reactants X and Y, respectively.

Solution ('integration') of the appropriate rate equation will tell us how the concentrations of reactants and products should vary over time.

For a simple **first-order reaction**:

$$d[X]/dt = -k[X]$$

the concentration of reactant, X, decreases exponentially with time:

$$[X] = [X]_0 . \exp(-kt)$$

where $[X]_0$ is the concentration at time zero.

Higher order reactions lead to more complicated expressions. However, for most of the biomolecular reactions studied, the kinetics are either first order (or simpler), or experimental conditions are arranged so as to make the process pseudo-first order. A **pseudo-first order** reaction is one which behaves as if it were first order in just one of the reactants (*e.g.* by keeping the concentrations of all other reactants constant or in excess).

A sometimes useful concept is the **half-life** or **half-time** ($t_{1/2}$) of a reaction, which is the time taken for the concentration of a given reactant to reach half its original value. For first-order reactions this is independent of absolute concentrations:

$$t_{1/2} = 0.693/k$$

We might ask: how fast can a reaction possibly go? In other words, if every molecular collision resulted in a reaction, how fast would that be? This is known as the **diffusion limit** and it provides a useful number with which to compare what we see in practice.

Molecules in solution move haphazardly under the influence of thermal motion. The rate at which they encounter one another will depend on their size and their diffusion coefficients. (See Section 4.5 for details of diffusion.) Large molecules will present larger targets for collision but will also tend to have smaller diffusion coefficients, and so will diffuse more slowly. This leads to an expression for the diffusion-controlled collision frequency:

$$A_{diffusion} = 4\pi N_A r_{XY}(D_X + D_Y) \times 1000$$

The factor of 1000 in this equation converts the SI units of volume ($m^3$) to $dm^3$.

where $D_X$ and $D_Y$ are the diffusion coefficients of the reacting molecules, and $r_{XY}$ is their **encounter distance**, giving the range within which they have to be for reaction to occur.

$A_{diffusion}$ would be the pre-exponential factor in the Arrhenius equation in the absence of molecular orientation, attraction or other effects. It provides an estimate of the upper limit of reaction rates under diffusional control with zero activation energy ($E_A = 0$).

**Worked Problem 6.1**

**Q**: Diffusion coefficients for small molecules in water at room temperature are about $1.5 \times 10^{-9}\,m^2\,s^{-1}$. What would be the diffusion controlled collision frequency assuming an encounter distance of $0.5\,nm$?

**A**: $A_{diffusion} = 4\pi N_A r_{XY}(D_X + D_Y) \times 1000$

$\qquad = 4\pi \times 6 \times 10^{23} \times 5 \times 10^{-10} \times 3 \times 10^{-9} \times 1000$

$\qquad \approx 1q10^{10}\,mol^{-1}\,dm^3\,s^{-1}$

Collision frequency $= A_{diffusion}\,[X][Y]$

**Worked Problem 6.2**

**Q**: What half-life would the above answer correspond to for a (pseudo) first-order reaction with reactant concentrations of $1 \times 10^{-3}\,mol\,dm^{-3}$?

**A**: Rate $k = A_{diffusion}\,[X] = 1 \times 10^{10} \times 1 \times 10^{-3} = 1 \times 10^7\,s^{-1}$

$t_{1/2} = 0.693/k \approx 0.693/1 \times 10^7 = 6.9 \times 10^{-8}\,s$ (69 ns)

Molecular diffusion in one or two dimensions is, in principle, faster than the three-dimensional diffusion calculated here. This has been used, for example, to explain the rapid kinetics of some reactions taking place on biological membranes (two-dimensional) or on DNA strands (one-dimensional diffusion).

Of course, in most living organisms, the reactions are taking place in a much more complicated environment than the laboratory flask. The cell is divided up into numerous compartments, reactions take place on membranes or adsorbed to other molecules or surfaces, and so forth. In such circumstances the equations describing kinetic processes become very complex. Nevertheless, the basic principles described here will still apply.

The experimental challenge is to devise techniques that reliably measure rate processes in biomolecules, usually under conditions pertinent to their biological function. For slower reactions, many of the traditional techniques of chemical kinetics can be used and reactions can be followed using a range of spectroscopic methods (described in earlier chapters). In the following sections we look at some of the more specialist methods that have been developed or adapted for biomolecular kinetics.

## 6.2    Rapid Reaction Techniques

Many reactions are too fast to follow by conventional mixing of reagents and, in any case, one often wishes to look at the very earliest stages of a reaction (**pre-steady state** kinetics). Such experiments require rapid and efficient mixing of reagents together with some means of following the subsequent reaction over a period of time.

### 6.2.1  Continuous Flow

One of the most straightforward ways of doing this is by continuous flow methods (Figure 6.2). Reagent solutions, X and Y, are pumped separately at constant flow rate into a mixing chamber from which the reacting mixture emerges and flows down the exit tube. The reaction time at any point downstream from the mixing chamber will depend on the combined flow rate and the dimensions of the exit tube. By positioning the detector at different positions along the tube or by varying the flow rate, the reaction can be monitored as a function of time. Detection might be by UV/visible absorbance, fluorescence or other suitable method.

---

**Worked Problem 6.3**

**Q**: For a combined flow rate of $10 \, cm^3 \, min^{-1}$, using tubing of 0.1 mm internal diameter, what reaction times could be followed with a flow tube 1–10 cm downstream from the mixing chamber?

**A**: Tube volume $= \pi r^2 l = \pi \times (0.005)^2 \times 1 = 7.8 \times 10^{-5} \, cm^3$ per cm

Volume flow rate, $10 \, cm^3 \, min^{-1} = 10\,000/60 = 0.167 \, cm^3 \, s^{-1}$

---

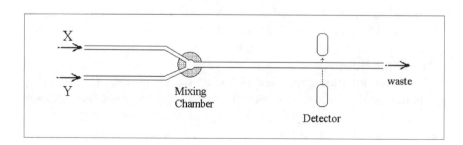

**Figure 6.2**  Sketch of continuous flow apparatus.

$$\text{Linear flow rate} = (7.8 \times 10^{-5})/0.167 = 4.7 \times 10^{-4}\,\text{s cm}^{-1}$$

$$\text{Thus: 1 cm downstream} \equiv 0.47\,\text{ms}$$

$$10\,\text{cm downstream} \equiv 4.7\,\text{ms}$$

One advantage of continuous flow methods is that the actual detection system can be quite unsophisticated since it merely has to monitor the steady concentration levels at a particular point in the flow tube. Reaction time is determined by the geometry and flow rate. One potential disadvantage, however, is that it can use up large quantities of reagent, especially at the rapid flow rates needed to cover the millisecond timescale.

## 6.2.2 Stopped-Flow

In stopped-flow kinetics, small volumes of reagent solutions are mixed rapidly in a flow cell by injection from separate syringes (Figure 6.3). The flow cell is typically fitted with transparent quartz windows so that UV/visible absorbance or fluorescence can be followed as a function of time. Circular dichroism (CD) kinetics can also be measured in some instruments. The injection pulse is controlled by a stopping syringe which, when it hits the end stop, electronically triggers the detector to start recording.

The typical '**dead time**' for most stopped flow instruments is around 1 ms, and is limited by the size of the flow cell and the efficiency of mixing. Reactions that are complete within this dead time cannot be measured, though more efficient mixing techniques are being developed to improve on this.

Stopped-flow devices of this kind have been used in many applications to study the initial steps in enzyme-catalysed reactions or ligand binding to biological macromolecules. In such experiments, one

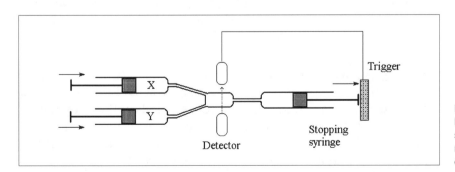

**Figure 6.3** Typical stopped-flow kinetics apparatus. The reservoir syringes and flow cell would normally be thermostatted to control the temperature.

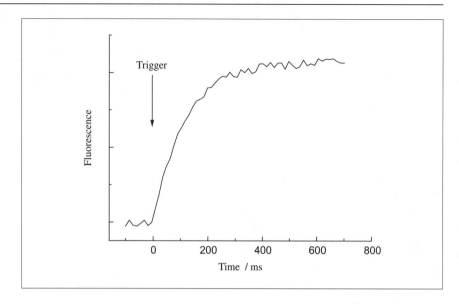

**Figure 6.4** Example of stopped-flow fluorescence changes for refolding of a globular protein in solution.

reservoir syringe would typically contain the enzyme or macro-molecule solution, and the other would contain substrate or ligand dissolved in the same buffer. The progress of the reaction over time (*e.g.* 1–1000 ms) would be followed using colour or UV absorbance changes as the substrate is converted into product, or fluorescence changes in the protein in response to ligand binding. In many cases of enzymic reactions, the reaction of interest might not have a convenient spectral change associated with it, but might be 'coupled' to a second step by addition of an excess of another enzyme or reagent mixture that can be used to monitor the primary reaction without disturbing its kinetics.

Stopped-flow methods are now commonly used to study the rates of folding or unfolding of proteins in solution by mixing with solutions that cause a pH jump or change in denaturant concentration. For example, proteins can be denatured (unfolded) by high concentrations (2–6 mol $dm^{-3}$) solutions of chemical denaturants such as urea or guanidinium hydrochloride. Stopped-flow dilution, by mixing with a large volume of buffer (using different size injection syringes), reverses this and allows the rapid kinetics of refolding to be followed (Figure 6.4).

## 6.3    Relaxation Methods

Not all reactions can be initiated by flow methods and the limitations imposed by inefficient mixing mean that much faster reactions are hard to study by such methods. An alternative approach is to take a mixture already at equilibrium, then rapidly perturb it, and watch the kinetics of its return to equilibrium. This is the basis for a number of **relaxation methods**.

Imagine a simple reaction:

$$A \underset{k_{-1}}{\overset{k_1}{\rightleftharpoons}} B$$

If the reaction is at equilibrium, $K = k_1/k_{-1} = [B]_{equilib}/[A]_{equilib}$ and $k_1[A]_{equilib} = k_{-1}[B]_{equilib}$.

If, however, the reaction is not quite at equilibrium, with concentration perturbed slightly:

$$[A] = [A]_{equilib} - [\delta A] \text{ and } [B] = [B]_{equilib} + [\delta B]$$

Then the return to equilibrium will be described by the rate equation:

$$\begin{aligned} d[A]/dt = d[\delta A]/dt &= -k_1[A] + k_2[B] \\ &= -k_1[A]_{equilib} + k_1[\delta A] + k_{-1}[B]_{equilib} + k_{-1}[\delta B] \\ &= (k_1 + k_{-1})[\delta A] \end{aligned}$$

since $k_1[A]_{equilib} = k_{-1}[B]_{equilib}$ and $[\delta B] = [\delta A]$ in this case.

This shows that the return to equilibrium after a small perturbation is first order, with a rate constant equal to $(k_1 + k_{-1})$. If the equilibrium constant, $K$, has been determined by other methods, then it is possible to obtain both the forward and reverse rate constants from this experiment.

Relaxation processes can be very fast. Measurements are limited only by the speed at which one can perturb the system and by how fast the detection system can follow the changes. Reaction kinetics on the μs timescale are normally feasible.

Various perturbation methods are used. Perhaps the most common is **temperature jump** (T-jump) in which the temperature of the sample is increased by a small amount (typically 5–10 °C) by passing a brief discharge of electric current through the solution, or by irradiation with a brief pulse of light from an infrared laser. **Pressure jump** (P-jump) perturbation can be done by first pressurizing the system using hydrostatic pressure (up to 1000 atm) and then releasing the pressure by breaking the seal. Some dye molecules change their affinity for hydrogen ions ($H^+$) when electronically excited, so absorption of light will bring about a change in pH of the solution. This can be used in **pH jump** experiments in which the sample solution, containing the dye molecule, is perturbed by brief irradiation from a pulsed laser at the appropriate wavelength to excite the dye.

This might, for example, represent the equilibrium between folded and unfolded states of a protein, which we might perturb by increasing the temperature.

### 6.3.1 Photochemical Reactions

For reactions that can be initiated by light, the start of the reaction (photon absorption) is essentially instantaneous. This means that the kinetics of such processes can be observed on much faster timescales without, for example, the practical limitations imposed by having to mix solutions. The technique, commonly called **flash photolysis**, involves exposure of the sample to a brief, intense pulse of light at an appropriate wavelength, following which the course reaction is followed by spectroscopic observation (absorbance or fluorescence).

These methods have been applied to studying various natural light-induced biomolecular processes such as photosynthesis and vision. With modern pulsed-laser techniques one can use light pulses as short as a few femtoseconds or less, and this gives information about the very earliest stages of important photochemical reactions.

$1\,fs = 10^{-15}\,s$

---

**Worked Problem 6.4**

**Q**: How far does light travel in 1 fs?

**A**: Speed of light (in vacuum) $= 3 \times 10^{10}\,m\,s^{-1}$

Distance in 1 fs $= 3 \times 10^{10} \times 10^{-15} = 3 \times 10^{-5}\,m = 30\,\mu m$

Note: the diameter of a human hair is about 50–100 μm

---

Flash photolysis methods can also be applied in other areas. For example, it has been found that intense pulses of laser light can displace carbon monoxide (CO) molecules bound to the haem groups in proteins such as haemoglobin and myoglobin. Once displaced, they migrate back to their original sites. This re-binding can be followed spectroscopically, giving information about the ways in which small molecules can diffuse through proteins.

Some dye molecules have a different $pK_a$ in the electronically excited state, so that they release or take up $H^+$ ions when exposed to a flash of light. This can be used to bring about rapid pH changes in solution. Such pH jump experiments can be used to follow the kinetics of biomolecular processes. Other kinds of photochemical reactions can be used to overcome mixing problems in rapid kinetic experiments. One example of this is the use of 'caged' ATP compounds that only become available for enzyme reaction, for example, when exposed to an intense light flash.[2]

## 6.4 Hydrogen Exchange

The hydrogen atoms on some groups are exchangeable and, using $D_2O$ for example, isotope exchange can be detected by techniques such as NMR, mass spectrometry, IR and Raman spectroscopy. Normally this exchange is very rapid (μs or less), but can be much slower if the group is protected from solvent by hydrogen bonding or burial within a macromolecular structure. This is particularly significant in the case of globular proteins, where amide hydrogen exchange of buried peptide groups can take days or longer. This can be exploited to gain information about the rates of conformational fluctuations in macromolecules.

A typical experiment involves mixing of the protein solution with isotopically enriched solvent (usually $D_2O$) under carefully controlled temperature and pH conditions, followed by sampling at timed intervals to monitor the extent of H–D exchange. Groups exposed to solvent at or near the surface of a globular protein will exchange rapidly, whereas those in more protected chemical environments will change more slowly.

The mechanism of slow hydrogen–deuterium exchange in these systems is quite complicated, but one simple model relates it to the rate of transient conformational (unfolding) transitions of the protein. During such conformational changes, groups that are normally protected become briefly exposed to solvent and can undergo isotope exchange.

Remember that proteins (and other biological macromolecules) are quite dynamic, flexible molecules. Even at equilibrium, and just like all chemical equilibrium processes, this equilibrium is dynamic, since thermal motion never stops. For a two-state process, the exchange kinetics may be described by:

$$N(H) \underset{k_{-1}}{\overset{k_1}{\rightleftharpoons}} U(H) \rightarrow U(D) \underset{D_2O}{\overset{k_{int}}{\rightleftharpoons}} N(D)$$

where N represents the 'native' or folded form of the protein, and U its unfolded state. $k_1$ and $k_{-1}$ are the forward and reverse rate constants for the unfolding transition, and $k_{int}$ is the intrinsic rate constant for hydrogen-deuterium exchange in a group fully exposed to solvent. The letters 'H' and 'D' indicate whether the protein group is labelled with hydrogen or deuterium, respectively.

Assuming that the $U \rightleftharpoons N$ process is rapid (as it would normally be), then the rate of isotope exchange for the folded protein would be:

$$-d[N(H)]/dt = d[N(D)]/dt = k_{int}.[U(H)]$$

Chemical groups such as –NH or –OH (but *not* –CH), involving hydrogens attached to electronegative atoms, rapidly (and reversibly) exchange to –ND or –OD when exposed to $D_2O$ (heavy water).

Alternatively, one might pre-equilibrate the protein molecules in a $D_2O$ buffer and then dilute into $H_2O$ to follow the release of exchangeable deuterium atoms. This can also be done using the radioactivity of tritium-labelled water.

'transient' = short-lived, momentary

In the steady state, $(k_{-1} + k_{int})[U(H)] = k_1[N(H)]$, so that:

$$-d[N(H)]/dt = d[N(D)]/dt = k_{int}.[N(H)] \times k_1/(k_{-1} + k_{int})$$

and the observed rate constant for isotope exchange:

$$k_{ex} = k_{int}.k_1/(k_{-1} + k_{int}).$$

Once a protein is unfolded (even transiently), then isotope exchange is likely to occur rapidly before it has chance to refold ($k_{int} > k_{-1}$), so $k_{ex} \approx k_1$. Under such conditions, measurement of isotope exchange rates in folded proteins can give information about the rates of naturally occurring conformational changes.

Of course, if this simple two-state model were entirely correct, then all exchangeable groups within a given protein would show the same exchange kinetics. Experimentally this is not usually the case, but there are numerous other exchange pathways that involve less extreme transient fluctuations in conformation that involve only partial unfolding of smaller regions of the protein.

## 6.5    Surface Plasmon Resonance

Surface plasmon resonance (SPR) is a relatively new technique that exploits some of the unusual physical properties of thin metallic films. When light reflects off a metal surface, a small proportion of the electromagnetic field penetrates into the metal where it interacts with the conduction electrons. In thin metal films this gives rise to collective motions of electrons, or **plasmons**, in the metal layer that can alter the reflective and other optical properties of the film (Figure 6.5). The magnitudes of these optical effects are very sensitive to the refractive index of the medium immediately in contact with the back face of the film. This is because the 'evanescent' or decaying electric field from the plasmon oscillations penetrates a short distance into the underlying medium (typically by about 300 nm).

At certain incident angles, the wavelength of the light matches the plasmon frequency and resonance occurs. This reduces the intensity of the reflected light; the magnitude of the effect depends on the refractive index of the material into which the evanescent wave penetrates.

This is the basis for SPR 'biosensor' chips for measuring the binding kinetics of biological molecules. The biosensor consists of a thin metal film (typically gold, deposited on a glass slide) to which proteins such as antibodies or other macromolecules may be attached (Figure 6.6).

Buffer, containing other molecules which may bind, flows continually over this surface at a fixed rate, usually $1–100\,\mu dm^3\,min^{-1}$. If any molecules in the solution bind to the immobilized molecules on the gold film, this produces a change in refractive index at the interface

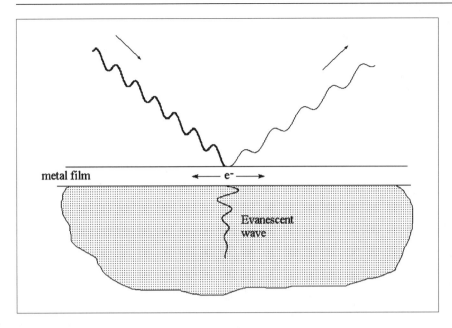

**Figure 6.5** Interaction of an electromagnetic wave with a thin metallic film.

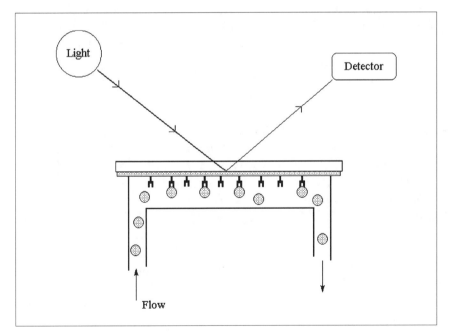

**Figure 6.6** Typical experimental arrangement for an SPR experiment to measure the binding of solute molecules (spheres) to immobilized receptors.

which is detected as a change in light intensity reflected from the film (Figure 6.7). This is usually measured in arbitrary **response units** (RU), which can be calibrated in terms of the mass of material bound per unit area of the chip.

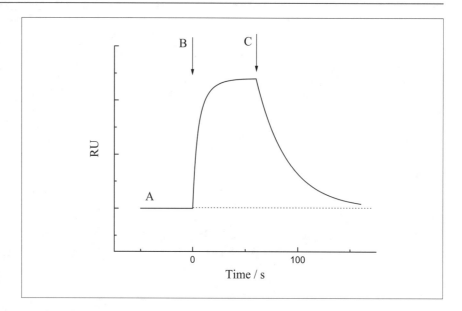

**Figure 6.7**   A typical sensorgram, recording RU as a function of time, illustrating the basic procedure.

Buffer solution is first flowed over the sensor chip to establish a blank baseline (A). Flow is then switched to the same buffer containing the test molecules (B). As these bind to the surface, the **sensorgram** response increases until an equilibrium level is reached. The shape of this binding or association phase can be analysed to give the kinetics of association ($k_{on}$) of the free and immobilized molecules. The flow may then be switched back to buffer (C) to wash off the (reversibly) bound molecules. The shape of this phase gives the kinetics of dissociation ($k_{off}$). For simple reversible binding processes, the ratio of $k_{on}$ to $k_{off}$ gives the equilibrium constant for the binding reaction, $K = k_{on}/k_{off}$, which can be related to thermodynamic properties in the usual way.

Various methods are available for attaching macromolecules or other ligands to the gold surfaces of sensor chips, either directly or by using other linker molecules. The advantage of SPR methods is that they can be very sensitive, and only very small amounts of material are needed to perform the measurements. Also, from a carefully controlled experiment, both kinetic and equilibrium parameters may be determined. A potential disadvantage is that immobilization of one of the components might affect its binding or other properties, and data must be interpreted with care.

## 6.6   Enzyme Kinetics

Enzymes are proteins that catalyze biochemical reactions with high specificity and efficiency. Enzymic processes can be quite complex, but the basic mechanism in all cases requires one or more discrete steps

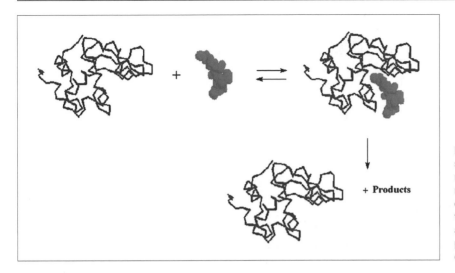

**Figure 6.8** Sketch of the basic steps in the Michaelis–Menten model in which the substrate molecule (black) forms an equilibrium complex by binding at the enzyme active site. Catalysis at the active site results in product formation and dissociation.

involving enzyme–substrate binding, followed by subsequent catalysis and release of reaction products (Figure 6.8). This means that the rate laws will be somewhat different from those normally found for simpler, non-catalysed reactions.

$$E + S \underset{k_{-1}}{\overset{k_1}{\rightleftharpoons}} ES \overset{k_2}{\longrightarrow} E + P$$

This is the classic Michaelis–Menten model of enzyme catalysis from which simple steady-state rate laws may be derived.

The rate of this reaction, v, that is the rate of formation of product, depends on the concentration of the enzyme–substrate complex, ES:

$$\text{Rate, } v = k_2[ES]$$

[ES] can be rewritten in other terms using the normal steady-state assumption:

$$d[ES]/dt = k_1[E][S] - (k_{-1} + k_2)[ES] = 0 \text{ in the steady state.}$$

If we define the total enzyme concentration, $C_E = [E] + [ES]$, this gives:

$$[ES] = C_E/(1 + K_M/[S])$$

where $K_M = (k_{-1} + k_2)/k_1$ is known as the **Michaelis constant**.

Consequently, the way in which the rate of the enzyme-catalysed reaction varies with substrate concentration should be:

$$v = k_2[ES] = k_2 C_E/(1 + K_M/[S]) \equiv v_{max}/(1 + K_M/[S])$$

where $v_{max} = k_2 C_E$ is the maximum rate of the reaction when all available enzyme active sites contain bound substrate ([ES] = $C_E$).

Leonor Michaelis (German–American biochemist, 1875–1949) and Maud Leonora Menten (Canadian physician and biochemist, 1879–1960) devised this model while working together in Berlin around 1912. Maud Menten was one of the first female doctors in Canada.

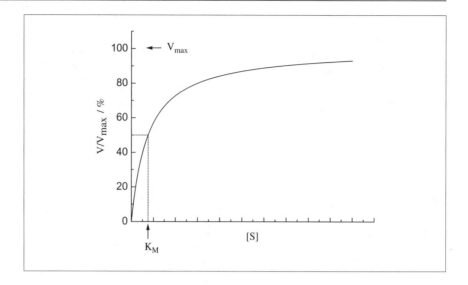

**Figure 6.9** Graph showing typical Michaelis–Menten behaviour for the change in reaction rate with increasing substrate concentration, [S].

Because enzymes are used at such low concentrations, it is usually OK to assume that the *free* substrate concentration, [S], is the same as the *total* substrate concentration.

This shows how enzyme-catalysed rates are expected to increase hyperbolically with substrate concentration (Figure 6.9). Although many enzymes do not necessarily follow this simple mechanism, the overall picture is nonetheless useful.

$K_M$ is a useful parameter that depends on both the enzyme and the substrate involved, as well as more general conditions like temperature, pH, ionic strength, *etc*. It can be conveniently visualized as the concentration of substrate that gives 50% of the maximum rate. It can also be related to the apparent binding affinity of the substrate to the enzyme. Since the **dissociation constant** for the enzyme–substrate complex formation is given by:

$$K_{diss} = [E][S]/[ES] = k_{-1}/k_1$$

it follows that:

$$K_M = K_{diss}(1 + k_2/k_1)$$

and $K_M \approx K_{diss}$ if $k_2 \ll k_1$

At low substrate concentrations, the rate appears pseudo-first order in [S]:

$$v \approx v_{max}[S]/K_M \,(\text{for } [S] \ll K_M)$$

whereas at very high substrate concentrations the process appears to be zero order, since it no longer changes with [S]:

$$v \approx v_{max} = k_2 C_E \,(\text{for } [S] \gg K_M)$$

The **specific activity** of the enzyme (sometimes designated $k_{cat}$) is $k_2 = v_{max}/C_E \equiv k_{cat}$ and the **catalytic efficiency** or **specificity constant** is indicated by the ratio, $k_{cat}/K_M$.

Generally speaking, catalytic efficiency is enhanced by tighter initial substrate binding (lower $K_M$) followed by a more rapid reaction of the ES complex (higher $k_2$).

## 6.6.1 Competitive Inhibition

Many natural and synthetic compounds (*e.g.* drugs) can compete with a substrate for binding at the enzyme active site, and this will inhibit the effectiveness of the enzyme. In simple competitive inhibition with inhibitor I, the enzyme may bind either I (to form EI) or S (to form ES), but never both together.

$$\textit{Either} \quad E + I \rightleftharpoons EI \quad \text{with } K_I = [E][I]/[EI]$$
$$\textit{or} \qquad E + S \rightleftharpoons ES \rightarrow E + P$$

Following the same steady-state method, but now with $C_E = [E] + [ES] + [EI]$, we get:

$$v = k_2 C_E / (1 + K_M^I / [S])$$

where $K_M^I = K_M (1 + [I]/K_I)$.

Thus the effect of adding a competitive inhibitor to the reaction mixture is to reduce the apparent binding affinity of the enzyme for substrate ($K_M^I > K_M$). Note, however, that $v_{max}$ is unaffected.

## 6.6.2 Other Kinds of Inhibition, Activation, Cooperativity and Allostery

There are numerous other ways in which the basic catalytic functions of enzymes can be altered by other molecules. Indeed, the ability to control the activities of enzymes is at least as important as the catalytic activity itself.

**Non-competitive inhibition**, in its simplest form, comes about when binding of an inhibitor affects the catalytic rate ($v_{max}$) rather than the binding affinity for substrate ($K_M$). This might happen, for instance, if the inhibitor molecule binds at some other site on the protein but induces some change at the active site that affects the catalytic activity. More generally, however, inhibitors can affect both $K_M$ and $v_{max}$.

It is equally important to be able to enhance enzymic rates under some circumstances, and numerous **activators** or **co-factors** behave in this way. Often, especially in multi-subunit enzymes, these effects can be **cooperative** or **allosteric**, meaning that the binding of effectors or inhibitors on one subunit can affect the binding or catalytic properties on another subunit. The mathematics of such processes can get very

The fastest Formula 1 racing car will never win a race without the ability to control its speed and direction.

complicated (and we won't do any of it here) but, in such cases, the simple Michaelis–Menten behaviour no longer applies. The advantage of such cooperative or allosteric effects is that they allow much tighter control of binding and catalytic processes.

### 6.6.3  Double Reciprocal Plots

Although computer software is now readily available to fit enzyme kinetic data to Michaelis–Menten and related equations, it can be instructive to use simple graphical methods in some cases. The most convenient of these (though not necessarily the most accurate) are based on double reciprocal methods that convert the hyperbolic rate equations into much simpler linear forms for plotting.

If we take the reciprocal of the Michaelis–Menten equation (*i.e.* turn both sides upside down) we get:

$$1/v = (1 + K_M/[S])/v_{max}$$

$$\text{or}: \quad 1/v = 1/v_{max} + K_M/[S]v_{max}$$

For historical reasons, such double reciprocal graphs are often called **Lineweaver–Burk** plots, after the two scientists who first devised its use.

This is the equation for a straight line plot of $1/v$ *versus* $1/[S]$ with intercept of $1/v_{max}$ and slope $K_M/v_{max}$. A similar plot in the presence of competitive inhibitor would also be linear, with the same intercept but greater slope (Figures 6.10 and 6.11).

A deviation from linear behaviour usually means that the simple Michaelis–Menten mechanism is not appropriate. Curved double reciprocal plots may indicate the presence of cooperative or allosteric effects in the enzyme.

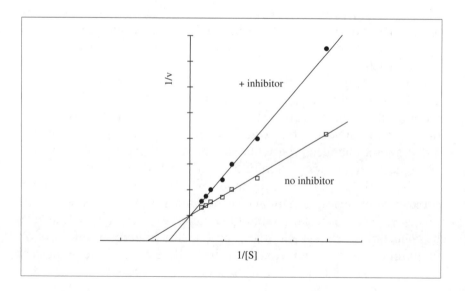

**Figure 6.10**  Double reciprocal plot for simple competitive inhibition.

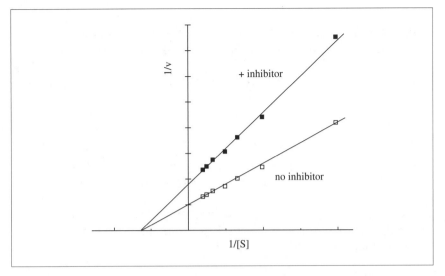

**Figure 6.11** Double reciprocal plot for simple non-competitive inhibition.

## Summary of Key Points

1. Biomolecular kinetics follow the same rules as elsewhere in chemistry.
2. Experimental methods for following fast reactions are based on rapid mixing, relaxation and photochemical effects.
3. Other techniques based on isotope exchange and plasmon resonance can also give information about protein dynamics and binding kinetics.
4. The kinetics of simple enzyme catalysis and inhibition can be analysed using the Michaelis–Menten model.

## Problems

**6.1.** The translational diffusion coefficient (D) for small globular proteins in water at room temperature is about $10^{-10} \, m^2 \, sec^{-1}$. What might be the diffusion controlled rate constant for encounter of such proteins with a membrane receptor or DNA target site with an encounter distance of 1 nm?

**6.2.** What would be the rate and $t_{1/2}$ for this process with a protein concentration of 1 μmolar $(1 \times 10^{-6} \, \text{mol dm}^{-3})$?

**6.3.** Sometimes reactions are found to have rate constants higher than the theoretical diffusion controlled rate. What might be some possible reasons for this?

**6.4.** The first step in the light-activated isomerization of retinal in the visual photoreceptor protein, rhodopsin, occurs in less than 6 ps. How far would a beam of light travel in this time?

**6.5.** Surface plasmon resonance experiments on the kinetics of binding of a peptide (in solution) to an immobilized receptor protein gave the following rate data:

| [peptide]/nmol dm$^{-3}$ | Rate of association/s$^{-1}$ |
| --- | --- |
| 1.2 | 0.023 |
| 3.6 | 0.068 |
| 4.8 | 0.091 |

a. Show that this is consistent with the association being first-order in peptide concentration.

b. What is the rate constant for association, $k_{on}$?

c. When the ligand solution was replaced by pure buffer solution, the half-time for dissociation of the bound peptide was found to be 7.2 s. What is the rate constant for dissociation, $k_{off}$?

d. What is the binding affinity constant for this peptide–protein complex?

## References

1. Discussion meeting issue, 'Quantum catalysis in enzymes—beyond the transition state theory paradigm' organized by Leslie Dutton, Nigel Scrutton, Mike Sutcliffe and Andrew Munro, *Phil. Trans. R. Soc. B.*, 2006, **361**, 1293–1455.
2. J. A. Dantzig, H. Higuchi and Y. E. Goldman, Studies of molecular motors using caged compounds, *Methods Enzymol.*, 1998, **291**, 307–334.

## Further Reading

A. Fersht, *Structure and Mechanism in Protein Science: A Guide to Enzyme Catalysis and Protein Folding*, Freeman, New York, 1999.

N. C. Price, R. A. Dwek, R. G. Ratcliffe and M. R. Wormald, *Physical Chemistry for Biochemists*, Oxford University Press, Oxford, 3rd edn, 2001, ch. 9–11.

I. Tinoco, K. Sauer, J. C. Wang and J. D. Puglisi, *Physical Chemistry: Principles and Applications in Biological Sciences*, Prentice Hall, Upper Saddle River, NJ, 4th edn, 2002, ch. 7–8.

# 7
# Chromatography and Electrophoresis

Chromatography and electrophoresis are general methods for characterizing and purifying molecules on the basis of their size, charge and other properties. They are widely used in research and industry for the analysis and preparation of biomolecules.

### Aims

You will probably be familiar with the analytical uses of paper chromatography or thin layer chromatography (TLC) from other branches of chemistry. Here we will consider mainly those aspects of chromatography and electrophoresis most relevant to biomolecules. After completing this chapter you should be able to:

- Describe the basic principles of chromatography and electrophoresis
- Explain how these techniques can be used to characterize biological molecules
- Describe some of their applications
- Make rational choices about the most suitable methods for separating molecules on the basis of their size, charge or other physico-chemical properties

## 7.1   Chromatography

IUPAC stands for the International Union of Pure and Applied Chemistry.

The IUPAC (1993) definition of chromatography is: 'Chromatography is a physical method of separation in which the components to be separated are distributed between two phases, one of which is stationary while the other moves in a definite direction.' Consequently, the basic arrangement for any chromatographic separation may be viewed as shown in Figure 7.1.

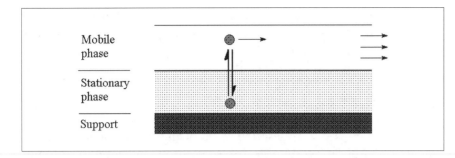

Mobile phase

Stationary phase

Support

**Figure 7.1** Chromatographic separation involves a supported stationary phase over which flows a mobile carrier phase. Solute molecules may move back and forth between stationary and mobile phases.

Within this general scheme, there is much flexibility in the kind of stationary and mobile phases that may be used, but the basic principles of separation remain the same. Consider a molecule dissolved in the mobile phase flowing over the stationary phase. If the solvent (mobile phase) is moving at a velocity, $v_s$, then the solute molecule will be carried along at the same speed. But if the solute molecule partitions into or binds to the stationary phase then, for the fraction of the time it spends in the stationary phase, it will remain stationary. Consequently, as the molecule hops back and forth between the two phases, its rate of flow will be reduced depending on how much time it spends in the stationary phase.

This can be quantified using ideas from phase equilibrium, rather like the solvent extraction or phase separation methods used in synthetic chemistry. The **partition** or **distribution coefficient** $K$, is defined as follows:

$$K = [\text{concentration of solute in mobile phase}]/$$
$$[\text{concentration of solute in the stationary phase}]$$

where the square brackets indicate equilibrium concentrations. In this dynamic equilibrium, the fraction of the time the molecule spends in the mobile phase will be $K/(1 + K)$. Consequently, the actual flow rate of the solute molecule will be:

$$v = v_S.K/(1 + K)$$

Different molecules, with different affinities or partition coefficients, will therefore move at different rates with respect to the solvent front. This is the basis for separation.

Chromatography can be performed on thin sheets, as in paper or thin-layer chromatography, in which case the sheets are often stained to show the relative positions of the separate compounds. Alternatively, and nowadays more frequently with biomolecules, the chromatographic medium can be packed into columns through which the mobile phase flows and the separated molecules are detected as they emerge (or 'elute') from the column.

The term 'chromatography' derives from the Greek *chroma* ('colour') and *graphein* ('writing'). It was first used by the Russian botanist Mikhail Semenovich Tsvett (1872–1919) to describe his separation of plant pigments on columns of finely divided calcium carbonate.

The 1948 Nobel Prize in Chemistry was awarded to the Swedish scientist Arne Tiselius '. . . for his research on electrophoresis and absorption analysis, especially for his discoveries concerning the complex nature of the serum proteins.'

The situation is rather like stepping on or off a moving conveyor belt: your actual rate of travel will depend on what fraction of the time you spend on the moving belt.

In paper or thin-layer chromatography, the absorbent paper or layer of finely divided silica acts as both a support and a stationary phase, also trapping stationary solvent as the bulk solvent passes through by capillary action. In gas chromatography (GC), the stationary phase is usually a thin layer of oil on the inner surface of a long capillary tube, through which is pumped an inert carrier gas (the mobile phase) carrying the volatile sample molecules.

The following sections describe some of the various methods of liquid chromatography suitable for separation and analysis of biological (macro)molecules. Such systems often use high pressures and rapid flow rates, and are sometimes loosely described as 'high performance liquid chromatography' (HPLC) or 'fast protein liquid chromatography' (FPLC).

### 7.1.1  Gel Filtration/Size Exclusion Chromatography

Gel filtration separates molecules in solution on the basis of their size (Figure 7.2). A gel filtration/size exclusion column is packed with a gel or slurry of fine particles of insoluble crosslinked carbohydrate materials (usually based on cellulose, agarose or dextran). These particles are carefully prepared so that they each contain a labyrinth of various size pores or cavities. The stationary phase, in this case, is simply the solvent buffer that is trapped within these cavities. The size range of the cavities is such that some macromolecules may enter, but larger ones are excluded. Consequently, as a solution of macromolecules is

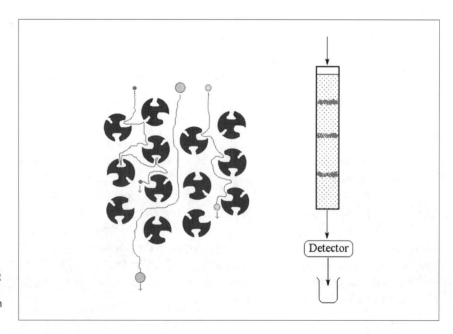

**Figure 7.2**  Size exclusion chromatography: molecules of different sizes will follow different paths through the matrix. Larger molecules follow the shorter path and so will emerge first.

pumped through the column, smaller molecules may follow a longer and more tortuous path through the matrix, whereas larger molecules flow through more rapidly.

Gel filtration/size exclusion columns can be calibrated by measuring the elution volumes of standard proteins (or other macromolecules) of known size. They may then be used to estimate sizes of unknown samples. Quite often this can be useful in determining whether a particular protein exists as a dimer, or other oligomer, in solution.

Although we might know the size of the monomer (from its amino acid or DNA sequence), this tells us nothing about its tertiary or quaternary structure, so experimental methods like gel filtration are needed here. We can also use these methods to detect binding between different molecules in solution, since they will tend to travel together down the column regardless of individual size.

## 7.1.2 Ion Exchange Chromatography

Ion exchange chromatography separates molecules on the basis of their charge. In ion exchange chromatography, the stationary phase is a matrix of chemically modified crosslinked polysaccharides or resins carrying a net positive (anion exchange) or negative (cation exchange) charge. The accompanying counter-ions (*e.g.* $Cl^-$ or $Na^+$) are free in solution as part of the mobile eluting buffer phase (Figure 7.3).

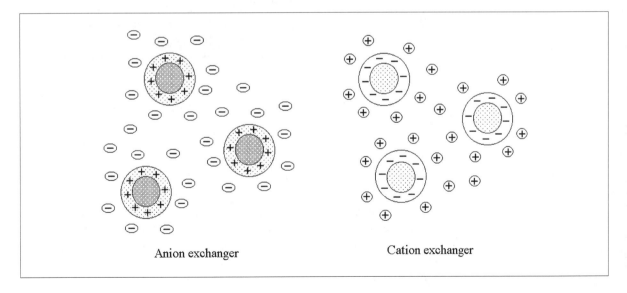

Anion exchanger                    Cation exchanger

**Figure 7.3**  Ion exchange beads with associated counter-ions.

When a solution of charged macromolecules is passed through such a column, they will tend to displace the small counter-ions (hence 'ion exchange') and stick to the column matrix by electrostatic interaction. In many cases the attachment is so strong that the macromolecule sticks firmly to the column material, and elution requires a higher ionic strength buffer (to weaken the electrostatic attraction by ionic screening) or a change in pH (to change the charge on the protein) to recover the sample. By careful optimization of buffer pH and concentrations, including salt- or pH-gradient elution, it is possible to separate quite complex mixtures.

The decision whether to use anion or cation exchange columns depends on the charge(s) on the molecules to be separated. This in turn depends, for proteins in particular, on the pH and the amino acid composition and structure of the protein.

You will recall from Chapter 1 (Section 1.8) that the isoelectric point (pI) is defined as the pH at which the protein molecule carries a net zero charge (equal numbers of positive or negatively charged groups). At this pH it would normally have a relatively poor attraction for an ion exchange medium. At a higher pH (pH > pI), $H^+$ ions will dissociate and the protein will become more negatively charged; this would tend to bind better to an anion exchange medium. Conversely, at lower pH (pH < pI), the protein will have a net positive charge and would bind better to a cation exchange column.

### 7.1.3  Affinity Chromatography

Affinity chromatography exploits the specific binding properties of macromolecules to separate on the basis of their affinity for particular groups or ligands (Figure 7.4). One example would be an antibody affinity column, containing immobilized antibodies specific for the molecule of interest. Other approaches might use small immobilized groups that resemble the substrate or receptor for specific enzymes or other proteins.

Many proteins are nowadays produced by recombinant DNA methods. Frequently these are prepared as 'fusion proteins' specifically designed to assist in purification by affinity chromatography. One common approach is to use 'histidine tagging', in which a short sequence of histidine residues (typically 6–10 His) is attached on the end of the polypeptide chain. In most cases this allows the protein to fold naturally but, at the same time, provides a means of isolating the protein on an affinity column that will specifically recognize the terminal histidine sequence. The imidazole side chain of histidine has a particular affinity for binding metal ions—nickel in particular. Nickel chelate columns, involving a stationary phase of nickel ions bound to an

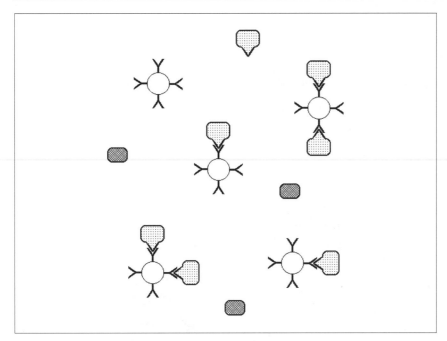

**Figure 7.4** Affinity chromatography matrix: functional groups or receptors on the stationary phase mimic the natural binding target for the chosen protein. Those proteins lacking this binding site will pass through the matrix unimpeded.

inert resin, will bind His-tagged proteins in a mixture. Proteins and other components in the mixture will wash straight through the column, and the required protein can then be recovered from the column either by changing the pH (which changes the metal ion binding affinity) or by displacement with free histidine or imidazole in the mobile buffer phase.

The term 'reversed phase' is used because, in conventional chromatography, the stationary phase is usually the more polar.

Many other strategies involving affinity techniques have been devised.

### 7.1.4  Reversed Phase Chromatography

Reversed phase chromatography separates molecules on the basis of their hydrophobicity or polarity.

The stationary phase here consists of an inert support material (usually silica) to which is attached long chain hydrocarbons (typically 8–18 –CH₂– groups). Hydrophobic molecules (in water) will tend to bind preferentially to this non-polar stationary phase, and mixtures will be separated according to their hydrophobicity (Figure 7.5). Often it is necessary to use non-polar solvent mixtures and solvent gradients (*e.g.* acetonitrile–water) for elution.

Reverse phase chromatography is often used for the analysis of peptides and other relatively small biomolecules. Other kinds of **hydrophobic interaction columns**, employing polysaccharide supports modified with less hydrophobic groups (phenyl groups, or shorter hydrocarbons), are used for protein separations.

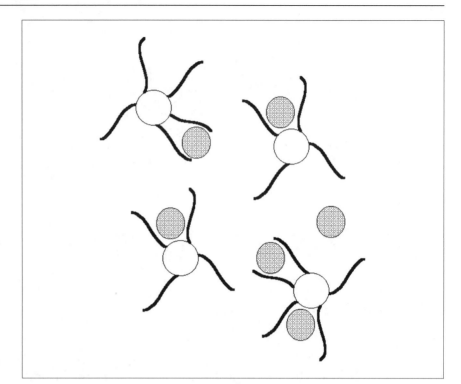

**Figure 7.5** Reverse phase chromatography matrix: hydrophobic molecules in solution (shaded) tend to associate with the non-polar hydrocarbon chains attached to the stationary phase.

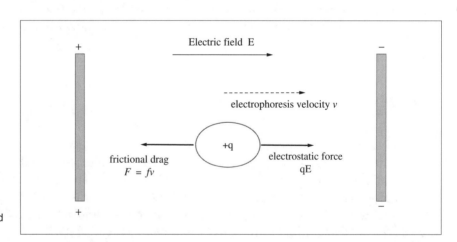

**Figure 7.6** Motion of a charged particle in an electric field.

## 7.2    Electrophoresis

Charged molecules in an electric field will tend to move towards the positive or negative electrode, depending on the sign of the molecular charge (Figure 7.6). For molecules in solution, this motion will be opposed by viscous drag from the surrounding environment, so the resulting **electrophoretic mobility** will depend on a number of factors including the overall charge, size and shape of the molecule. This is the principle of electrophoresis.

Using the same ideas that we introduced earlier when discussing **sedimentation rates** (Section 4.4), in the steady state for a particle with charge $+q$ in an electric field ($E$), the electrostatic force on the particle will be exactly balanced by the viscous drag, so that:

$$fv = qE$$

and the electrophoretic velocity of the particle, $v = qE/f$. Where $f$ is the **frictional coefficient** defined earlier (Section 4.4). In reality the situation is a little more complicated than this because of the movement of counter-ions and other electrolytes. However, this does illustrate the basic principle that bigger charge or smaller size (lower $f$) will lead to more rapid electrophoretic migration.

The term 'electrophoresis' can be roughly translated to mean 'charge migration'. It derives from the Greek words *electron* ('amber') and *phoresis* ('carrying'). Amber is a naturally occurring insulating material, formed from the fossilized sap or resin from ancient trees, which was known to easily acquire a static electric charge when rubbed.

### 7.2.1  Gel Electrophoresis

For practical purposes, most electrophoresis of biological molecules is performed in **gels** rather than just using the solution. This cuts down on the diffusion and convection that would interfere with sharp separations in liquids. It also makes it easier to stain and detect the samples at the end of the experiment (Figure 7.7).

The rectangular slab gel is normally just a few millimetres thick and is made of crosslinked polymeric materials such as polyacrylamide or agarose. The gel is mainly water (buffer), containing just a few percent (5–15%) of the polymer material. However this is sufficient to give a relatively rigid gel (a stiff jelly) that will retain the sample after the experiment, but yet allows relatively free passage of macromolecules (and buffer) through the gel matrix.

The sample bands are not usually visible during electrophoresis (unless intrinsically coloured or fluorescent), but are normally visualized afterwards by immersing the gel in specific staining solutions. For DNA samples, intercalating fluorescent dyes such as ethidium bromide (Section 2.4) are used, whereas proteins can be visualized using histological stains such as Coomassie blue, or the more sensitive silver stains.

Proteins come in all shapes and sizes, and their charges (both sign and magnitude) will depend on the particular protein and the pH used. Consequently there is usually no way to predict where a particular protein band might appear in this kind of electrophoresis (but see SDS-PAGE, Section 7.2.2).

Nucleic acids are a little simpler. DNA molecules are generally rod-shaped, with a net negative charge (arising from the phosphate

**Capillary electrophoresis** can be used in some cases. Here the solutions are contained in very fine capillaries where convection and diffusion are less of a problem, thus eliminating the need for a gel matrix.

Gel electrophoresis of DNA fragments is a major tool in determining gene sequences. Sanger's 1980 Nobel Prize in chemistry (his second) and the subsequent elucidation of human and other genomes depended on this.

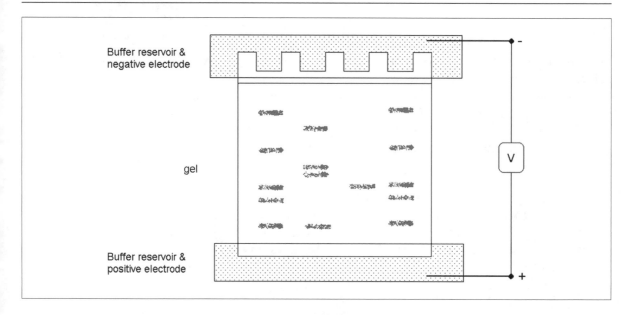

Buffer reservoir &
negative electrode

gel

V

Buffer reservoir &
positive electrode

**Figure 7.7** Slab gel electrophoresis. Each end of the gel is immersed in an aqueous buffer solution that also acts as an electrode and allows application of a high voltage (V) across the gel (top to bottom in this diagram). Samples are injected as small volumes into the rectangular 'wells' at the top of the gel. When the voltage is applied, negatively charged molecules (in this case) migrate into the gel and move down as different 'bands', depending on their electrophoretic mobility. Several sample 'lanes' (four in this diagram) are generally used to allow direct comparison of several samples and standards.

SDS = sodium dodecyl sulfate $[Me(CH_2)_{10}CH_2OSO_2^- Na^+]$, also known as sodium lauryl sulfate, is a common component of household detergents and shampoos; PAGE = polyacrylamide gel electrophoresis.

backbone) that depends only the length of the chain. Consequently, it is found that the electrophoretic mobility of DNA is just proportional to the length of the chain. DNA fragments differing by as little as one nucleotide can be seen as separate, adjacent bands, and this has been used extensively for DNA sequencing. Anything that perturbs this simple relationship between mobility and chain length (*e.g.* a change in conformation or binding of another molecule) will show up as a shift in the normal band position. This has been used to develop **band-shift assays** to measure DNA–protein and other interactions by electrophoresis.

### 7.2.2 SDS-PAGE

The lack of information regarding protein size in normal (non-denaturing) electrophoresis (previous section) can be overcome by denaturing (unfolding) the protein in a strong detergent solution. Sodium dodecyl sulfate is a strong detergent that disrupts the native structure of proteins in solution and forms a micelle encapsulating the unfolded polypeptide chain (Figure 7.8).

This micelle usually contains a mass ratio of SDS to protein of around 1.4 : 1, corresponding to roughly 0.5 detergent molecules per peptide group. Consequently, these micellar particles have a size and (negative) charge roughly in proportion to the size of the original protein molecule, and will therefore migrate accordingly in electrophoresis.

SDS denaturation of proteins for SDS-PAGE is normally carried out by heating in the presence of sulfydryl reducing agents in order to

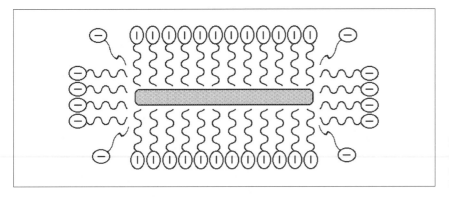

**Figure 7.8** Idealized view of the clustering of SDS molecules around a denatured protein molecule (shaded).

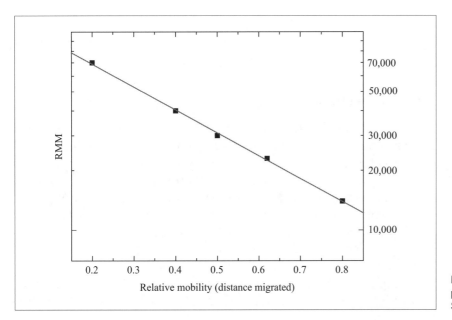

**Figure 7.9** Relative mobility of proteins of different sizes on SDS-PAGE.

break any disulfide crosslinks in the protein structure. Subsequent electrophoresis uses the same arrangement as described in the previous section, but with the addition of SDS to the electrophoresis buffer (Figure 7.7).

It is found experimentally that the electrophoretic mobility under these conditions is roughly proportional to the logarithm of the protein RMM (Figure 7.9); this can be used to estimate the apparent size of an unknown protein.[1] However, this generally only gives a rough estimate of the size of the polypeptide chains and is dependent upon the validity of the assumptions regarding SDS binding. Different families of proteins, especially those with significant chemical modifications, crosslinks, glycosylation, *etc.*, show deviations from this simple behaviour.

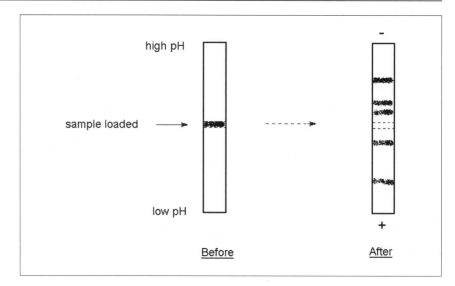

high pH

sample loaded  ⟶

low pH

Before                    After

**Figure 7.10** Electrophoresis in a pH-gradient: isoelectric focusing of a mixture of proteins with different pIs.

An **ampholyte** is a substance that can act as either an acid or a base. Amino acids, for example, containing both acidic (–COOH) and basic (–NH$_2$) groups are ampholytes. A range of pI values can be obtained by varying the ratio of acidic and basic groups in short polymers.

## 7.2.3  Isoelectric Focusing

Isoelectric focusing (IEF) is electrophoresis using a pH gradient. Proteins (and other molecules) will only move in an electric field if they carry a net electrical charge. This charge is zero when the pH of the solution equals the pI (isoelectric point). Protein molecules undergoing electrophoresis in an isoelectric focusing gel will migrate through the pH-gradient gel until they reach their pI. At this point they have zero net charge and will migrate no further (Figure 7.10).

The pH gradient in the gel is created by using proprietary mixtures of small ampholytes with good buffering capacity.

During electrophoresis, the ampholytes will themselves migrate to their pI in the gel and, because of their buffering capacity, will establish the pH gradient. Alternatively, the ampholyte groups may be covalently attached or immobilized in the gel matrix, and the gradient formed by carefully pouring a gradated mixture of components before the gel sets.

## 7.2.4  Two-dimensional Gel Electrophoresis

It is possible to combine different methods of electrophoresis to give much better resolution of complex mixtures of proteins. For example, it has been estimated that the number of different proteins in an individual cell may be between 5000 (bacteria) and 50 000 (human). Many of these will have similar size or charge or pI, and resolution of so many bands on a single gel would be impossible. Two-dimensional (2-D) electrophoresis combines the techniques of SDS-PAGE and isoelectric focusing (IEF) to give improved separation (Figure 7.11).

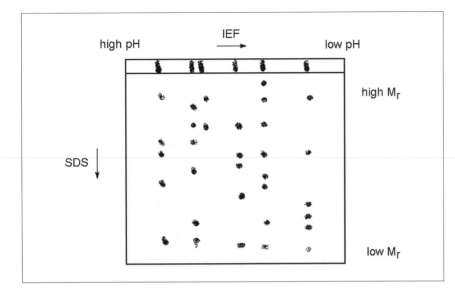

high pH     IEF →     low pH

SDS ↓

high $M_r$

low $M_r$

**Figure 7.11** Two-dimensional electrophoresis of a complex mixture of different proteins. In this hypothetical example, only six different bands are seen in the first dimension (IEF). But when further separated on the basis of size (SDS, second dimension), many more different protein bands appear.

The protein mixture is first separated by IEF (first dimension). This gel is then laid alongside a rectangular SDS slab gel and electrophoresis is run perpendicular to the original direction. Proteins are therefore separated in two dimensions on the basis of both their apparent size (SDS-PAGE) and pI (IEF). Several thousand different proteins can often be resolved in this way.

2-D gel electrophoresis is currently a major tool in the field of **proteomics** which, by analogy with **genomics**, is attempting to map all the proteins in a living organism. See also Chapter 3.

## Summary of Key Points

1. Chromatography is the separation of molecules according to the way they partition between stationary and mobile phases.
2. Different chromatography methods may separate molecules on the basis of their size, charge, polarity, binding affinity or a combination of such factors.
3. Migration of molecules in an electric field (electrophoresis) depends on the charge, size and shape of the migrating particles.

## Problems

**7.1.** A research team has used recombinant DNA methods to clone and express an unusual protein from a parasitic worm. This protein has a sequence of histidine residues at its *N*-terminal end and, for laboratory purposes, is produced in bacterial cells. How might this protein be purified from a mixture of other proteins in the bacterial expression system?

**7.2.** The DNA and amino acid sequence gives a relative molecular mass of around 15 000 for the polypeptide chain of the protein in Problem 7.1, but it was initially thought that the protein might occur as a dimer in the native state. How might this be checked chromatographically?

**7.3.** Could electrophoresis methods be used to help resolve this question (Problem 7.2)?

**7.4.** The natural protein from the worm (Problem 7.1) is known to bind long-chain fatty acids, but the recombinant protein from the laboratory was initially found to have less binding affinity than expected. One possibility was that the recombinant protein site was already occupied by fatty acids from the bacterial cell system. How might such contaminants be removed?

## References

1. K. Weber, J. R. Pringle and M. Osborn, Measurement of molecular weights by electrophoresis on SDS-acrylamide gel, *Methods Enzymol.*, 1972, **26**, 3–27.

## Further Reading

D. Sheehan, *Physical Biochemistry: Principles and Applications*, Wiley, New York, 2nd edn, 2009, ch. 2, 5.

I. Tinoco, K. Sauer, J. C. Wang and J. D. Puglisi, *Physical Chemistry: Principles and Applications in Biological Sciences*, Prentice Hall, Upper Saddle River, NJ, 4th edn, 2002, ch. 6.

# 8
# Imaging

Science would be so much easier if we could actually see what the atoms and molecules looked like. So what's stopping us? In this chapter (and the next) we take a look at the ways in which we might visualize the structures of objects at the microscopic level and beyond.

## Aims

Here we examine the various possible ways of obtaining images of microscopic objects using electromagnetic waves and particles. This is a big topic and we shall only be able to skim the surface, but after working through this section you should be able to:

- Understand how reconstruction of an image from scattered waves requires both amplitude and phase information.
- Be aware of how diffraction effects can limit the resolution of optical instruments
- Describe the basics of X-ray scattering and crystallographic structure determination
- Outline the elements of protein crystallography and the phase problem
- Appreciate how neutrons and electrons may also be used for imaging

## 8.1 Waves and Particles

Most of us are fortunate enough to have the ability to 'see with our own eyes' and interpret the objects and the world around us in terms of these mental pictures. But what is actually happening in this process and can we apply it to much smaller objects?

Forming an image of something usually involves the scattering, reflection or transmission of electromagnetic radiation ('light'), and many imaging processes both depend upon and are limited by the wavelike properties of such radiation. We should also be aware that,

Wave–particle duality was first expressed quantitatively by Louis de Broglie in his equation, $\lambda = h/mv$, which relates the wave-like behaviour of any particle to its momentum where $\lambda$ is the wavelength, h is Planck's constant, $m$ is the particle's mass at rest and $v$ is the particle's velocity (*i.e.* momentum = $mv$).

Velocity (c), wavelength ($\lambda$) and frequency (f) are linked by the fundamental relation: $c = f\lambda$. Amplitude and phase are independent variables.

Waves are **monochromatic** ('single-coloured') if they have the same frequency (or wavelength).

because of quantum mechanical wave–particle duality, things that we might normally think of as particles (electrons, neutrons, *etc.*) can also behave like waves under certain circumstances. We will see below how this might be exploited for imaging. But first we need to review some of the characteristics of wavelike motion.

Wavelike motion in any medium is characterized by wavelength, frequency, velocity, amplitude and phase. The concept of 'phase' is crucial in this chapter since the relative phases of different waves determine how they might combine to form images. The phase difference between two waves can be viewed as the amount by which one wave lags behind the other, either in time or space (Figure 8.1). Waves are said to be 'in phase' when their peaks and troughs coincide. Two waves are said to be 'out of phase' when the peaks of one coincide with the troughs of the other.

Waves radiate in all directions from a point (*e.g.* ripples on a pond). This gives rise to diffraction and interference effects when waves scattered from different points combine or recombine. Common phenomena such as diffraction patterns or interference fringes, and colours in thin films depend on the relative phases of scattered light.

If two waves (of the same frequency) are in phase when they meet, they will add together to give a wave whose amplitude is the sum of the two. This is known as **constructive interference** (Figure 8.2). Conversely, if the waves are out of phase when they meet, the positive peaks of one will be exactly cancelled by the negative peaks of the other and *vice versa*. This is **destructive interference.**

## 8.2    Lenses or No Lenses: Reconstructing the Image

To form an image of any object using electromagnetic radiation ('light'), we need some way of recombining (some of) the scattered

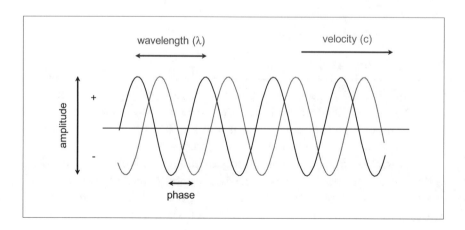

**Figure 8.1** Waves are characterized by their wavelength, velocity, amplitude and phase.

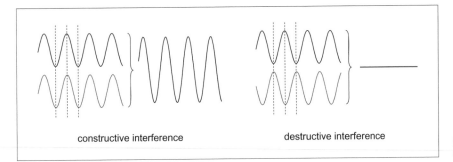

constructive interference              destructive interference

**Figure 8.2** Constructive and destructive interference.

light in such a way that the scattered waves are **in phase** with each other when they get back together. This is what lenses and mirrors do when producing focused images using light in the visible, ultraviolet (UV) or infrared (IR) region.

An important concept in optics is **Fermat's principle of least time.** This states that, of all the possible paths that might be taken between two points, light takes the path which requires the least time. This can be used to design all sorts of optical instruments. In the case of image formation using a simple glass lens (Figure 8.3), since glass has a higher refractive index than air, light scattered by the object (A) is slowed down when passing through the lens. The shape of the lens is such that lots of different least-time paths from A to the image point (B) are possible. Furthermore, since all of these different paths take the same (least) time, all the waves will have the same phase at B and will therefore recombine constructively to form the image.

### 8.2.1  Diffraction and the Limits of Resolution

With optical microscopes, magnification is achieved by using combinations of powerful lenses to get ever closer views of the sample. But what are the limits? Why can't we just add more lenses to get even closer views, possibly down to the molecular level? The problem is that typical molecular dimensions (0.5–5 nm) are very much smaller than the wavelengths of visible light (around 500 nm). This is well below the diffraction limit that determines the ultimate resolving power of microscopes. Visible images are formed using scattered light. However, waves exhibit diffraction and interference effects, and these become much more significant when the dimensions of the object are comparable to the wavelength of light, or smaller. Below this **diffraction limit** it is not possible to resolve individual objects by conventional optics.

Imagine trying to see two closely separated dots (points) through a microscope. The image produced at the focal point is not perfect but is

**Refractive index**. The speed of light is constant in a vacuum ($c = 3 \times 10^8\,m\,s^{-1}$) regardless of wavelength, but electromagnetic waves are slowed down to different extents when travelling through transparent substances. This gives rise to **refraction**: the bending of light beams when they pass from one transparent medium to another. The refractive index ($n$) is defined as the ratio, $n$ = (speed of light in the medium)/(speed of light in vacuum). Typical values at visible wavelengths are 1.0003 (air), 1.33 (water), 1.4–1.8 (glasses, various), 1.4 (eye lens).

In practice, no simple lens can be perfect. Real lenses are subject to various aberrations, since it is difficult to design the exact shape that precisely satisfies Fermat's principle, especially over a range of wavelengths (since refractive index varies with wavelength).

By analogy: ocean waves are blocked by a large object such as an island, but they pass around a small object such as a pebble as if it wasn't there.

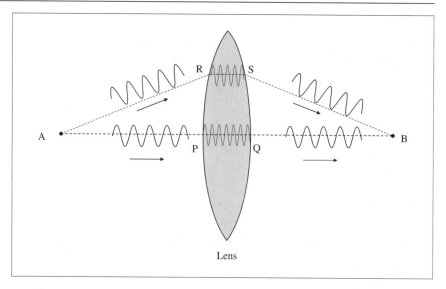

**Figure 8.3** How a lens works. The shape of the perfect lens is such that, when refractive index is taken into account, the time taken for any ray to travel from the object (A) to the image point (B) is the same, regardless of path. For example, although path ARSB is longer than APQB, the light spends less time in the slow lane (*i.e.* the lens), so all the waves are back in phase at B.

rather the fuzzy diffraction pattern of the dots. If these fuzzy patterns overlap too much, then we are not able to see the dots as separate objects. This depends on the wavelength of the light and the size of the apertures in the system, and this puts a limit on the resolving power of any microscope.

The **Rayleigh criterion** is that two images are just resolved if the central maximum of the diffraction pattern of one object is at the first minimum in the diffraction pattern of the second object. This gives an estimate of resolving power, $\phi \approx 1.22\lambda/d$, where $\phi$ is the angular separation of the objects, $\lambda$ is the wavelength, and $d$ is the diameter of the optical aperture.

For the human eye, with a pupil diameter of about 2.5 mm in normal daylight, and light in the middle of the visible region (550 nm), this resolving power corresponds to a separation of about 0.1 mm at normal reading distance. In other words, with good eyesight, in good light, we should be able to see two dots separated on this page by about 0.1–0.2 mm. This is borne out by experiment. Optical microscopes are subject to the same limitations, but effectively bring the image closer to our eyes so that we can resolve more closely spaced objects. For good microscopes at high magnifications ($\sim 600 \times$) the theoretical resolvable separation approaches the wavelength of the light, but is usually worse in practice because of lens aberrations and practical limits on aperture sizes and sample illumination.

Emerging technologies that might extend the resolution of optical microscopy beyond the diffraction limit are discussed in refs. 1–3.

So why not just use shorter wavelength light to get improved resolution and see even smaller objects in a conventional microscope? The problems are practical. Although glass and, more specifically, quartz can be used for lenses into the UV region (maybe down to 150 nm), most substances absorb light at shorter wavelengths and cannot be used to construct appropriate lenses. Alternative approaches are needed.

## 8.2.2 Confocal and other Microscopies

Another way of constructing an image is to scan over the object with a beam of light and measure the amount of light scattered (or transmitted) at each point. This is the basis for many confocal and related microscopy techniques, and differs from conventional microscopy where the entire sample is illuminated and imaged through the microscope lenses. In a typical device, a sharply focused laser beam is scanned systematically across the entire sample, measuring the intensity of light scattered at each point. Different layers in the sample can be sampled by changing the focal point of the scanning beam (or the sample position), and the resulting intensities can be combined electronically to give a two- or three-dimensional representation of the object. A crucial element of a confocal microscope is the pinhole aperture placed at the equivalent focal point in front of the photodetector (Figure 8.4), which blocks out-of-focus light coming from other depths in the sample.

The term 'confocal' arises because the pinhole aperture is *con*jugate to the *focal* point of the lens in this optical configuration.

Confocal microscopy most commonly uses samples stained with fluorescent dyes to enhance image contrast and to highlight specific

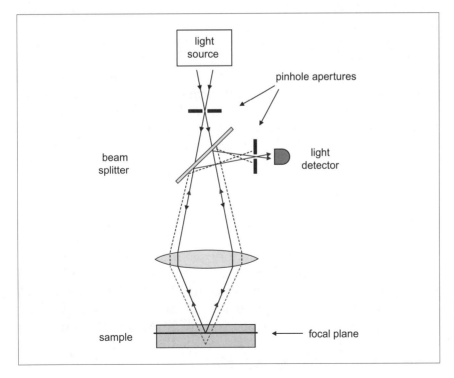

**Figure 8.4** Optical configuration of a confocal microscope. Note how out-of-focus light (dashed lines) is blocked by the confocal detector aperture.

features (Section 2.4.5). This is often coupled with two-photon excitation (Section 2.4.10).

The resolution of confocal microscopes and similar optical configurations is determined ultimately by the size of the scanning spot, and this is subject to the same diffraction limits as conventional microscopy (Section 8.2.1). It is impossible to produce an infinitely sharp point of light; diffraction will always give fuzzy spots when dimensions approach the wavelength of the light.

## 8.3     X-ray Diffraction and Protein Crystallography

Diffraction and interference effects limit the resolution of all optical instruments, and we cannot usually see detail in objects smaller than the wavelength of the light we are using (Section 8.2.1). So, in order to see atomic or molecular detail, we need to use wavelengths of atomic or molecular dimensions—in other words, wavelengths of order 0.1 pm (1 Å). For electromagnetic radiation, this means X-rays. However, there are no lenses or mirrors capable of focusing beams of X-rays sufficiently well to make X-ray microscopes, and we need a different approach. Fortunately diffraction, which is a nuisance in optical microscopy, comes to our aid here. This section describes the basics of diffraction techniques, concentrating on methods used to determine the crystal structures of proteins and other biological macromolecules.

Methods for producing protein crystals are described in Section 5.8.4.

The principles are relatively simple. A monochromatic beam of X-rays is focused onto the sample—a crystal in this case. The X-rays are diffracted by the crystal, giving rise to a pattern of spots whose intensity and position depend on the symmetry in the crystal and the structure of the molecules in the crystal lattice. This information is then used to calculate the electron density distribution (*i.e.* molecular structure) in the crystal. That's where the fun starts . . .

### 8.3.1   Generation of X-rays

Historically, X-rays were first observed in experiments where high-energy electrons were focused onto a metal target, and this remains the most common method for laboratory and medical applications. In an X-ray tube, electrons from a heated wire filament are accelerated in a vacuum using high voltages (40 000 to 60 000 volts) and focused onto a solid copper target. The rapid deceleration and variety of collision processes gives rise to a broad spectrum of X-rays together with more intense X-ray emission at specific wavelengths corresponding to electronic transitions in the target atoms.

Similar principles are used in old-fashioned 'cathode ray tubes' and TV screens in which the electrons are focused onto the 'phosphor' to create a spot of light.

Synchrotron radiation is now much more commonly used for high-resolution biomolecular structure determination. The circular trajectories of particles (usually electrons) moving at speeds close to the velocity of light in high-energy accelerators ('synchrotrons') gives rise to intense beams of electromagnetic radiation covering the IR to X-ray range.

For copper, the most intense X-ray emissions are at 0.15418 pm (1.5418 Å), known as K$\alpha$, and 0.13922 pm (1.3922 Å) known as K$\beta$. Specific wavelengths can be selected by using thin sheets of different metals as selective absorbance filters, or by using different metal targets in the X-ray tube.

### 8.3.2  Detection of X-rays

In a diffraction or scattering experiment, we need to be able to measure the **intensity** of X-rays scattered in a particular direction. Historically X-rays were first detected by their effects on photographic film or plates, and this remained the method of choice for many years. However, film requires time-consuming chemical processing, and the amount of darkening on the developed film is only proportional to the intensity of the diffracted X-rays over a restricted range. Nowadays film has been mostly replaced by semiconductor charge-coupled devices (CCD, see Section 2.2.1) similar to those used in most digital cameras. These have a much better dynamic range and can capture a two-dimensional map of the intensity of X-rays scattered in chosen directions.

Although initially a fortuitous by-product of high-energy nuclear physics research, **synchrotron radiation** is now of such importance in biomolecular structure, materials science and chemistry that dedicated national and international facilities have now been built, for example the Diamond Light Source in the UK.

### 8.3.3  Scattering of X-rays

X-rays in the range of interest here (0.5–1.5 Å) are scattered predominantly by electrons in the sample, and the amount of scattering depends on the electron density in the sample.

Elastic (Thompson) scattering occurs without change in energy of the X-rays and is the form of scattering most relevant here. Electrons in the sample oscillate in response to the oscillating electric field of the incident X-rays. Acting like tiny radio antennas (see Section 2.1), they will re-radiate electromagnetic waves of the same frequency, but no longer in the same direction as the original beam. Such scattering is normally coherent, *i.e.* in phase with the incident beam. This is critically important for diffraction.

The X-ray scattering power of an atom depends on the number of electrons in the atom and on the direction of scatter ($\theta$), and is determined by the **atomic scattering factor**, $f$. This is greatest in the forward direction ($\theta = 0$) and decreases at higher scattering angles (see Figure 8.5). Note that $f$ has units of electrons and equals the number of electrons in the atom at $\theta = 0$.

X-rays may also be scattered inelastically, *i.e.* with loss of energy. **Compton scattering** involves interaction with outer shell electrons in which sufficient energy is transferred to the electron to eject it from its

This is the basis for medical imaging applications of X-rays. Water and soft biological tissue has relatively low electron density; typically more than 90% of X-rays will pass through unchanged. Hard tissues such as teeth and bone have higher electron densities because they contain calcium and other heavy elements, and will cast more of a shadow.

**Thompson scattering** may also be viewed quantum mechanically as arising from a virtual transition to a disallowed state.

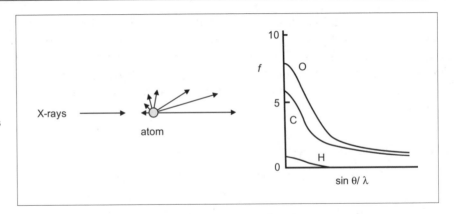

**Figure 8.5** Angular dependence of the atomic scattering factor. X-rays are scattered with different intensities in different directions (predominantly in the forward direction, $\theta = 0$) as indicated by the vectors (arrows), or graphically in terms of the scattering angle, $\theta$).

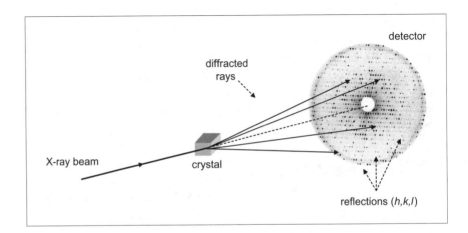

**Figure 8.6** X-ray diffraction from a single crystal.

atomic or molecular orbital. The scattered X-ray is incoherent (*i.e.* out of phase) and of lower energy (longer wavelength). Interaction with inner shell electrons can give rise to photoelectric absorption, again with ejection of the electron but with residual energy usually dissipated as heat. These processes can result in significant radiation damage to samples during exposure to X-rays.

### 8.3.4 The Diffraction Experiment

When a beam of X-rays is focused on a crystal, some of the X-rays scatter in specific directions to give a characteristic pattern of spots (or 'reflections') on the detector (Figure 8.6). These spots arise by diffraction from different crystal lattice planes and rotation of the crystal allows collection of the complete diffraction pattern for the electron density in the crystal.

## Box 8.1 Bragg's Law

Each spot in the diffraction pattern can be visualized as coming from constructive interference of X-rays reflected from the atoms (or molecules) lying in planes of the crystal lattice. For adjacent planes, separated by distance $d$, constructive interference between reflected waves will only occur at angles ($\theta$) where the phase difference (distance ABC in Figure 8.7) corresponds to an integral number of wavelengths:

$$\text{Distance ABC} = 2d.\sin\theta = n\lambda$$

This is Bragg's Law, named after W. L. Bragg who first proposed it in 1912.

Note that each plane can give rise to multiple reflections: n = 1 (first order), n = 2 (second order), and so forth, though the intensity usually decreases for higher orders.

Note also that smaller lattice spacings (smaller $d$) give larger diffraction angles ($2\theta$) and *vice versa*. This means that. for higher resolution information, it is necessary to collect data to larger diffraction angles.

Father and son, W. H. (William Henry) and W. L. (William Lawrence) Bragg were jointly awarded the Nobel Prize in Physics in 1915 for their work on X-ray diffraction from crystals.

It will be useful here to revise what you may have learnt elsewhere about symmetry and crystal lattices. See also Further Reading.

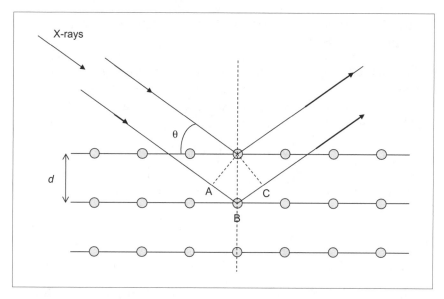

**Figure 8.7**  Bragg's law diffraction from a simple crystal. Note that the beam is reflected by an angle $2\theta$.

The spacings of the diffraction spots and the symmetry of the diffraction pattern depend on the lattice spacing, unit cell dimensions and symmetry of the crystal. Each reflection corresponds to constructive interference from a particular set of crystallographic planes, and can

For any crystal it is possible to draw numerous planes in different directions through the crystal lattice. The Miller index, *h k l*, is a way of describing each plane uniquely. It is named after the British mineralogist, William Hallowes Miller, who introduced this classification scheme in the early 19th century. It is closely related to the concept of the **reciprocal lattice**. Each set of planes in the real crystal lattice corresponds to a point (*h,k,l*) in the reciprocal lattice.

be labelled (or 'indexed') using three numbers (integers), *h k l*, called **Miller indices**.

From the symmetry and spacing of the spots in a diffraction pattern, one can tell things about the symmetry of the diffracting crystal and the size and shape of the unit cell. But they say nothing about the contents of the unit cell—and that's what we would like to see.

However, information about the electron density of the atoms and molecules in the crystal is given by the strength or intensities of the diffraction spots. The important quantity here is something called the **structure factor**. This is defined mathematically in Box 8.2, but, by analogy with the atomic scattering factor (Section 8.3.3), can be viewed as describing the wave scattered by the combination of atoms in the unit cell. The intensity, I(*h,k,l*), of any reflection is related to the square of the amplitude of the structure factor, F(*h,k,l*), by:

$$I(h, k, l) = a.|F(h, k, l)|^2$$

where *a* is a numerical constant that takes account of geometric factors and absorption corrections, and $|F(h,k,l)|$ is the modulus or amplitude of the structure factor.

---

### Box 8.2 Structure Factors, Electron Density and Fourier Transforms

For any reflection (*h,k,l*) the structure factor, F(*h,k,l*), is given by summation of all the atomic scattering factors for the atoms in the unit cell, corrected by a phase factor that accounts for the relative positions of the atoms in the unit cell:

$$F(h, k, l) = \Sigma f(j).\exp[2\pi i(hx(j) + ky(j) + lz(j)] \qquad (8.1)$$

where $f(j)$ is the scattering factor of atom $j$, with coordinates [$x(j)$, $y(j)$, $z(j)$], and the summation ($\Sigma$) is taken over all the atoms in the unit cell.

The **electron density distribution**, $\rho(x,y,z)$, is related to the structure factors as follows:

$$\rho(x, y, z) = (1/V).\Sigma\Sigma\Sigma F(h, k, l).\exp[-2\pi i(hx + ky + lz)] \quad (8.2)$$

where $V$ is the volume of the unit cell and the (triple) summation ($\Sigma\Sigma\Sigma$) is taken over all possible combinations of *h,k,l*.

Note that eqn (8.1) and eqn (8.2) are known in mathematics as Fourier transforms. The electron density distribution is the Fourier transform of the structure factors, and *vice versa*.

I($h,k,l$) is what we measure experimentally, so this allows us to calculate $|F(h,k,l)|$ for each reflection. Since the structure factor is just the Fourier transform of the electron density (see Box 8.2), straightforward mathematical manipulation should ideally give us the structure (electron density). Unfortunately, however, $F(h,k,l)$ is a complex, wavelike quantity that has not only amplitude but phase, and we need the phase of each reflection, $\phi(h,k,l)$ in order to complete the calculation.

Fourier transforms, named after the French mathematician Jean Baptiste Joseph Fourier, are incredibly useful mathematical tools with wide applications. In general, the diffraction pattern of any object is described by its Fourier transform.

### 8.3.5 The Phase Problem—and How to Solve it

To obtain the electron density distribution from single crystal diffraction data we need both the amplitudes, $|F(h,k,l)|$, and the phases, $\phi(h,k,l)$, of each of the reflections in order to perform the reverse Fourier transform that converts $F(h,k,l)$ into $\rho(x,y,z)$. We know the amplitudes from the intensities of the diffraction spots, but detectors can measure only the magnitude of the scattered X-ray, not its phase, so phase information is lost in the measurement. This requires additional experiments, and several methods have been devised for proteins and other biological macromolecules.

One of the earliest methods used was to form **heavy metal derivatives**, following and extending the **isomorphous replacement** procedures developed for small molecule crystallography. Heavy metals such as mercury can be adsorbed at specific protein binding sites by soaking the crystals in a suitable salt solution. The heavy metal, being more electron-dense, gives much stronger X-ray scattering, contributing measurably to each structure factor (eqn 8.1), with subtle changes in resultant phase and amplitude. By comparing the diffraction data of the heavy metal derivative with that of the original crystal, and assuming that the metal binding doesn't change the protein structure too much, the position(s) of the heavy metal atoms within the crystal can be determined. From this, estimates of the phase angles for each structure factor can be obtained. In fact, analysis of diffraction data from just one derivative gives a pair of possible phases, $\phi(h,k,l)$, for each reflection, and it is usually necessary to use two or more different derivatives to resolve the ambiguities. However, heavy atoms absorb X-rays in a wavelength specific manner, leading to changes in both the amplitude and phase of the resultant scattering and this **anomalous dispersion** can be used to resolve the phase ambiguity.

The term **isomorphous** derives from the Greek, meaning 'same shape'. Heavy atom and isomorphous replacement methods for solving the phase problem in chemical crystallography were first developed by John Monteath Robertson (Gardiner Professor of Chemistry at Glasgow University, 1942–1970). In 1939 he predicted that the structure of insulin would be solved in this way–a task finally accomplished by Dorothy Hodgkin and her Oxford group in 1969.

More recent methods take advantage of recombinant DNA methods that allow the expression and purification of proteins incorporating non-standard amino acids. One particular approach introduces selenomethionine in place of the normal methionine in the bacterial growth medium. This results in complete incorporation of a heavy atom into specific positions within the expressed protein (replacing methionine S by Se in this case)—usually with little effect on protein structure. The resulting protein can then be crystallized and diffraction data collected as before. However, the ability to accurately change the X-ray wavelength from a synchrotron source means that the full potential of anomalous dispersion can be exploited for solving the phase problem. The **multiple-wavelength anomalous dispersion** (MAD) method simply requires that accurate data be collected from the same crystal, but using different X-ray wavelengths corresponding to the peak and inflection point of the heavy atom absorption edge, together with further data at wavelengths remote from the absorption edge. This gives sufficient information to predict the positions of the heavy atoms and thus unambiguously solve the phase problem. This is currently the method of choice for solving new protein structures.

In cases where potentially similar molecular structures have been determined (or calculated), it is possible to use **molecular replacement methods** as a starting point for calculating phases. The previously determined atomic coordinates of a (supposedly) similar protein are used in a rotational and translational search process to try to match the new data and hence give an estimate of phases. Iterative computational methods can then be used to improve the agreement between the molecular model and the X-ray data.

Selenomethionine (2-amino-4-methylselenyl-butanoic acid) is an α-amino acid with a –$CH_2$–$CH_2$–Se–$CH_3$ side chain, compared to –$CH_2$–$CH_2$–S–$CH_3$ in methionine.

### 8.3.6  Building the Structure

The end product of any successful X-ray diffraction experiment is an **electron density map**. This is usually displayed as a set of contour lines in two- or three-dimensions, representing the electron density distribution in the crystallographic unit cell (see, for example, Figure 8.8). The amount of detail (or 'resolution') visible in this map will depend on the quality of the original diffraction data, and it is at this stage that chemical intuition and experience are often needed to resolve ambiguities and construct a sensible structural model of the molecule(s) in the unit cell. Remember that this might include water, salts and other components in the crystal growth medium, as well as the macromolecules themselves. Water molecules and buffer ions can sometimes be resolved, but more often do not bind at specific locations and appear more as background noise in the electron density. In most cases (for proteins, at least) the amino acid sequence will be known, and it is usually possible to use this to locate and trace the electron density of the polypeptide backbone. Side chains are often less well resolved. Structure refinement programs are often used at this stage to ensure that the stereochemistry of the structural model is satisfactory and consistent with the electron density data.

It is easy to be seduced at this stage by the impressive quality of computer graphic images of biomolecular structures. Always try to bear in mind that these are (usually) static cartoons and that the pictures can only be as good as the original experimental data.

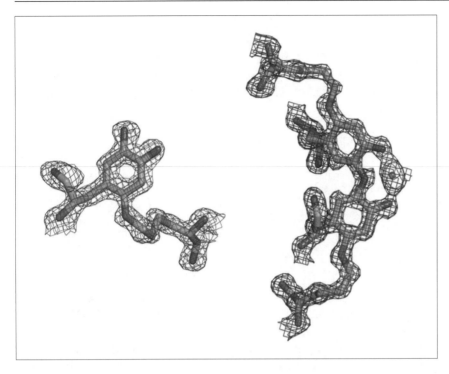

**Figure 8.8** Example of portions of an electron density map derived from high resolution X-ray diffraction from a protein crystal. The wire-mesh grid shows the experimental data (electron density contours) to which the molecular model can be fitted.

## 8.4  Fibre Diffraction and Small-angle Scattering

Less well-ordered samples can also show diffraction effects that can be used to give useful structural information. For example, two-dimensional ordering of molecules in fibres can give X-ray diffraction patterns containing information about lateral spacing of molecules in the fibre, or molecular order along the length of the fibre.

A famous example of this is the original X-ray diffraction pattern of moist DNA fibres, determined by Rosalind Franklin and her associate, R. G. Gosling, in 1953.[4] This showed Bragg reflections consistent with the 3.4 Å base stacking distance, together with diffraction at smaller angles corresponding to a 34 Å periodicity, in agreement with the Watson and Crick double-helix model of DNA that was published simultaneously.[5] Small-angle fibre diffraction can give information about molecular packing on a larger scale.

Remember from Bragg's law (Box 8.1) that smaller angle diffraction corresponds to longer-range periodicity (larger lattice spacing), and *vice versa*.

### Worked Problem 8.1

**Q:** Connective tissue (skin, bone, tendon, *etc.*) is made up of collagen fibres in which tropocollagen molecules stack side-by-side. Each tropocollagen molecule is a long, thin triple helical protein molecule, about 3000 Å long and 15 Å in diameter. X-ray diffraction from oriented collagen fibres gives small angle reflections along the fibre axis corresponding to 0.138° with an X-ray wavelength of 1.5418 Å. What lattice spacing does this suggest?

**A:** Use Bragg's law (Box 8.1)

$$2d.\sin\theta = n\lambda$$

Scattering angle $2\theta = 0.138; \theta = 0.069°$

Spacing $d = n\lambda/2.\sin\theta$

$$= 1.5418/(2.\sin 0.069), i.e. \text{ assuming first order, n} = 1$$

$$= 640\,\text{Å}(64\,\text{nm})$$

Note that this spacing bears no obvious correspondence with either the molecular diameter (15 Å) or length (3000 Å). It actually represents the periodicity of the staggered arrangement of tropocollagen molecules in the fibre. The molecules do not line up side-by-side, but rather are displaced longitudinally along the fibre by multiples of 640 Å. This periodicity can also be seen clearly in electron microscope images of connective tissues.

Small-angle scattering from macromolecules in solution can give low-resolution information about molecular size and shape. However, detailed diffraction patterns of individual molecules are smeared out by random motions and rotations of the molecules in solution, and there is no way to recover phase information. The approach usually adopted is to compare experimental X-ray scattering curves (intensity *versus* scattering angle) to theoretical expectations for model structures of different shapes and sizes.

## 8.5    Neutron Diffraction and Scattering

Diffraction and scattering effects can also be observed using beams of neutrons of appropriate kinetic energies. Thermal neutrons, with a kinetic energy equivalent to thermal motion at room temperature (Section 5.1), have an average velocity of about $2700\,\text{m s}^{-1}$ and a de Broglie wavelength of about 0.15 nm (1.5 Å). This means that the neutrons will behave like waves at the atomic and molecular level, and can be used just like X-rays to determine structural information from diffraction and scattering experiments.

Unlike X-rays, which interact predominantly with electron clouds around atoms, neutron scattering comes mainly from interactions with the atomic nuclei. Neutrons are neutral (uncharged) particles, so their paths are relatively unaffected by electrostatic effects but they are strongly affected by nuclear forces when they get close to an atomic nucleus. Different nuclei have different neutron scattering characteristics. Interaction with the hydrogen nucleus, for example, is particularly strong. This contrasts with X-ray scattering, where hydrogen atoms are often very hard to resolve because of their low electron density. This means that neutron diffraction experiments can be used to give more detailed information about hydrogen atom positions that can only usually be inferred indirectly from X-ray data.

Neutron scattering is also very sensitive to isotope effects. For example, neutron scattering from deuterium nuclei (one proton + one neutron) is significantly different from that of hydrogen (one proton). Consequently, hydrogen–deuterium exchange experiments (Section 6.4) show large effects with neutrons that are invisible with X-rays. By mixing different proportions of regular water ($H_2O$) and heavy water ($D_2O$), it is possible to obtain buffer solutions that are effectively transparent to neutrons, so only the sample molecules show up. This 'contrast matching' technique is particularly useful in small-angle neutron scattering experiments where different sample features can be enhanced against solvent background.

Neutron beams for research purposes are usually produced either by the 'fission' (splitting) of uranium atoms in a nuclear reactor, or by nuclear 'spallation' reactions when a high energy beam of protons is fired at a metal target. Please don't try either of these at home!

## 8.6 Electron Microscopy

Another way of forming an image (of sorts) is to look at the shadow cast on a screen by a beam of light (or particles). The shadow gives a two-dimensional outline representation of the opaque regions of the object, and if the illuminating beam is divergent, then the shadow image will be magnified.

But particles can act as waves and we can focus beams of electrons using electromagnetic fields. This forms the basis for electron microscopy, where beams of electrons (in a vacuum) are used to produce focused images of small objects.

### Worked Problem 8.2

**Q**: What is the de Broglie wavelength for a 100 keV electron?

**A**: The de Broglie wavelength for a particle of mass $m$ travelling at velocity $v$ is given by:

$$\lambda = h/mv, \text{ where } h \text{ is the Planck constant } (6.626 \times 10^{-34}\,\text{Js})$$

Kinetic energy, $E = \frac{1}{2}mv^2 = (mv)^2/2m$, so that we can write the momentum as:

$$mv = (2mE)^{1/2} \text{ and } \lambda = h/(2mE)^{1/2}$$

A kinetic energy of 1 eV corresponds to $1.6 \times 10^{-19}\,\text{J}$, so in this case:

$$E = 100\,\text{keV} = 1.6 \times 10^{-14}\,\text{J}$$

$$m = 9.1 \times 10^{-31}\,\text{kg for an electron.}$$

Hence $\quad \lambda = (6.626 \times 10^{-34})/(2 \times 9.1 \times 10^{-31} \times 1.6 \times 10^{-14})^{\frac{1}{2}}$
$= 3.9 \times 10^{-12}\,\text{m}$

$$\equiv 3.9\,\text{pm} \equiv 0.0039\,\text{nm}.$$

This is much smaller than a typical molecule, so such electron beams could be used for conventional imaging provided suitable lenses were available for focusing.

The electron gun and electromagnetic focusing lenses are similar in principle to those found in a conventional TV picture tube.

The transmission electron microscope is rather like a slide or movie projector, but using electrons instead of light (Figure 8.9). The beam of electrons is focused through the sample (the 'slide') and projected onto a large screen; magnifications up to $100\,000 \times$ or more can be obtained. The sample must be very thin, so as to allow at least some of the electrons to pass through, and everything is under high vacuum, otherwise the electron beam would be scattered by the air. The image obtained represents differences in electron scattering power of different regions of the sample.

The major technical drawbacks in using electron microscopy to image single (macro)molecules are **contrast** and **radiation damage.** The electron density in the sample molecule is usually very similar to the substrate or support film upon which it is mounted, so it can be hard to visualize against the background. Contrast can be enhanced by using heavy metal stains (with higher electron density), but this also tends to reduce the resolution of the image. Exposure of the sample to

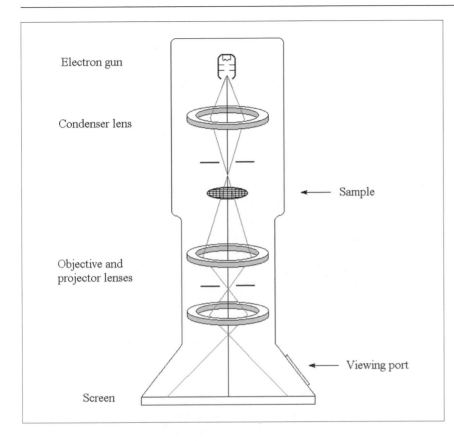

Electron gun

Condenser lens

Sample

Objective and
projector lenses

Viewing port

Screen

**Figure 8.9** Basic layout of a transmission electron microscope. Monochromatic electrons from an 'electron gun' are focused through an aperture by the electromagnetic condenser lens onto the sample. Electrons passing through the sample are collected and focused by the electromagnetic objective and projector lenses to form a magnified image on the viewing screen. Images can be captured using photographic plates or electronic devices.

the electron beam also gives rise to damage caused by absorption of some of the energetic electrons. This can be reduced by cooling the sample stage (**cryo-electron microscopy**) and by using short exposure times with lower beam currents. Despite such precautions, it is rarely possible to get a clear image of a single molecule and the most successful methods superimpose images from a large number of separate observations, using image reconstruction techniques to generate an overall picture of the molecule.

The surface structure of thicker samples can be examined using **scanning electron microscopy** (SEM). Here a fine beam of electrons is scanned across the surface of the specimen and the scattered secondary electrons emitted from the surface of the sample are detected electronically. Secondary electrons are best produced by electron collision with conduction electrons in a metal surface (by analogy with the photoelectric effect), so samples for SEM are usually coated by evaporation with a thin film of gold in order to make them more visible. This naturally limits the fine detail that can be seen.

Be careful not to confuse SEM (scanning electron microscopy) with STM (scanning tunnelling microscopy—see Section 9.3).

**Summary of Key Points**

1. Formation of images from scattered light requires waves to be recombined in phase.
2. This can be done automatically in optical microscopes for objects with dimensions down to a few hundred nanometres, but not smaller because of diffraction limits and lack of suitable lenses.
3. Diffraction from crystals at shorter (X-ray) wavelengths allows scattered waves to be recombined mathematically, provided the phases are known. This is the basis of protein crystallographic structure determination.
4. Similar methods can exploit the wavelike properties of neutrons (neutron diffraction and scattering) or the much shorter de Broglie wavelength of high energy electron beams (electron microscopy).

**Problems**

**8.1.** Haemoglobin and myoglobin are important proteins involved in oxygen binding, transport and exchange, and their crystal structures were among the first to be determined by X-ray diffraction methods in the 1950s and 1960s. Max Perutz, one of the scientists who shared the Nobel Prize for this work, often pointed out that there is no obvious path in these protein structures for oxygen molecules to get in or out of the $O_2$ binding sites. Does this mean that the structures are wrong? Discuss.

**8.2.** Rosalind Franklin's original X-ray diffraction experiments on moist DNA fibres (*ca.* 1952) showed strong first-order Bragg reflections at about 26.2°, with an X-ray wavelength of 1.5418 Å. What lattice spacing might this correspond to? What might this tell us about DNA molecular structure?

**8.3.** Calculate the mean velocity and de Broglie wavelength for a thermal neutron at 20 °C (Neutron mass $= 1.67 \times 10^{-27}$ kg).

## References

1. E. H. K. Stolzer, Light microscopy beyond the diffraction limit?, *Nature*, 2002, **417**, 806–807.
2. S. W. Hell, Far-field optical nanoscopy, *Science*, 2007, **316**, 1153–1158.
3. P. J. Walla, *Modern Biophysical Chemistry: Detection and Analysis of Biomolecules*, Wiley-VCH, Weinheim, 2009, ch. 7.
4. R. Franklin and R. G. Gosling, Molecular configuration in sodium thymonucleate, *Nature*, 1953, **171**, 740–741.
5. J. D. Watson and F. H. Crick, Molecular structure of nucleic acids; a structure for deoxyribose nucleic acid, *Nature*, 1953, **171**, 737–738.

## Further Reading

J. P. Glusker and K. N. Trueblood, *Crystal Structure Analysis—A Primer*, Oxford University Press, Oxford, 2nd edn, 1985.

# 9
# Single Molecules

We rarely get the opportunity to see how individual molecules behave. Mostly what we see are the average effects of large numbers of molecules. Molecular structures determined by crystallographic or NMR methods are time- and number-averages of a large population of molecules. Yet the behaviour of a living cell may be the result of just single molecule interactions; for example, the binding of just one repressor protein to a single DNA site can affect the entire behaviour of the cell. How might we study this? Molecules are always in motion and it is very hard to pin down individuals. This chapter covers some of the methods now being developed to tackle this problem.

---

**Aims**

After working through this chapter you should be able to:

- Estimate the numbers of molecules present in various circumstances
- Appreciate the fluctuating nature of single molecules
- List some of the methods used to study single molecules
- Describe the basic principles of atomic force microscopy and optical trapping
- Discuss single molecule fluorescence

## 9.1 How Many Molecules can Stand on the Head of a Pin?

Theologians used to speculate about how many angels might stand on the head of a pin. But, since few people had ever seen an angel, it was difficult to resolve the question. Few people have ever seen a single molecule either. Can we count them? The following worked examples should remind you how we do this.

---

**Worked Problem 9.1**

**Q**: How many molecules are there in $1\,cm^3$ of a $1\,mg\,cm^{-3}$ solution of macromolecules of relative molecular mass $50\,000$?

---

**A**: $1\,\text{mg} \equiv (1 \times 10^{-3}\,\text{g})/50\,000 = 2 \times 10^{-8}$ moles $\times\, 6 \times 10^{23} = 1.2 \times 10^{16}$ molecules per $\text{cm}^3$.

---

### Worked Problem 9.2

**Q**: How many molecules might be found in 1 cubic micron of such a solution?

**A**: 1 cubic micron $= (1 \times 10^{-4})^3\,\text{cm}^3 = 1 \times 10^{-12}\,\text{cm}^3$

This would contain $1 \times 10^{-12}\,\text{cm}^3 \times 1.2 \times 10^{16}$ molecules $\text{cm}^{-3}$ $= 12\,000$ molecules.

---

### Worked Problem 9.3

**Q**: Imagine a typical small bacterial cell of diameter around 1 micron (1 μm). What would be the effective molar concentration of a particular compound if there were just one such molecule per cell?

**A**: Approximate volume of the cell (since we don't know its exact shape) will be of order $(\text{diameter})^3 = 1 \times 10^{-18}$ $\text{m}^3 = 1 \times 10^{-15}\,\text{dm}^3$ $(1\,\text{m}^3 = 1000\,\text{dm}^3)$.

One molecule in $1 \times 10^{-15}\,\text{dm}^3$ corresponds to $1/(1 \times 10^{-15})$ $= 10^{15}$ *molecules* per $\text{dm}^3 = 10^{15}/N_A = 10^{15}/(6 \times 10^{23}) = 1.7 \times 10^{-9}$ *moles* per $\text{dm}^3 = 1.7\,\text{nmol}\,\text{dm}^{-3}$.

The volume of a *cube* (of side d) $= \text{d}^3$. The volume of a *sphere* (of diameter d) $= (4/3)\pi\text{r}^3 = (\pi/6)$ $\text{d}^3 \approx \frac{1}{2}\,\text{d}^3$.

---

### Worked Problem 9.4

**Q**: A typical small globular protein is about 2 nm in diameter. How many such molecules might pack side-by-side on:

(a) a 1 cm square postage stamp;

(b) the head of a pin, 1 mm across;

(c) the fine point of a very sharp pin, 10 nm across?

**A**: Assuming a square, two-dimensional array, each molecule will occupy an area of about $2 \times 2 = 4\,nm^2 \equiv 4 \times 10^{-14}\,cm^2$. Consequently . . .

(a) The postage stamp can hold $1/(4 \times 10^{-14}) = 2.5 \times 10^{13}$ molecules.

(b) The pin head can hold $0.1 \times 0.1/(4 \times 10^{-14}) = 2.5 \times 10^{11}$ molecules.

(c) The pin point can hold $10^{-6} \times 10^{-6}/(4 \times 10^{-14}) = 25$ molecules.

### Worked Problem 9.5

**Q**: As we saw in Chapter 1, the approximately three billion bases in the human genome would extend about 1 metre if stretched out end-to-end. What might be the diameter if this were coiled up as tightly as possible into a single ball?

**A**: Very roughly, assuming that each DNA base has a relative molecular mass of around 300, the RMM of the entire genome is about $300 \times 3 \times 10^9 = 9 \times 10^{11}\,g\,mole^{-1}$, or about $1.5 \times 10^{-12}\,g$ per molecule.

The density of organic matter is roughly $1.4\,g\,cm^{-3}$ (compared to $1.0\,g\,cm^{-3}$ for water), so one genome molecule will occupy about $10^{-12}\,cm^3$ (remember, we are only interested in round figures). This corresponds to a cube of side $10^{-4}\,cm$ (1 micron).

Typical animal cells may be 10–100 microns. The cell nucleus (which contains the genomic DNA) can be 5–20 $\mu$m across, depending on circumstances.

## 9.2    Thermodynamic Fluctuations and the Ergodic Hypothesis

Atoms and molecules are never at rest (except at absolute 0 K). Everything at the molecular level is in a constant state of turmoil—moving, rotating, vibrating and colliding with neighbours under the influence of thermal motion. This is what we call **heat.** For normal, everyday objects this turmoil is imperceptible. The things we feel such as hotness, coldness, pressure, *etc.* are a consequence of these molecular motions, but we don't feel the individual fluctuations because they all average out among the vast numbers of molecules involved. What is it like at the single molecule level?

There is, of course, quantum mechanical 'zero point energy' at 0 K, but this is not motion in the classical sense.

---

**Worked Problem 9.6**

**Q**: The average thermal kinetic energy of any object is $3kT/2$. What is the root mean square (r.m.s.) velocity of translational motion of a macromolecule of RMM 25 000 at 37 °C?

**A**: Mean kinetic energy $\frac{1}{2}m\langle v^2\rangle = 3kT/2$; therefore, r.m.s. velocity, $\langle v^2\rangle^{1/2} = (3kT/m)^{1/2}$.
For a single macromolecule, $m = 25\,000/N_A = 25\,000/6 \times 10^{23} = 4.2 \times 10^{-20}\,\text{g} \equiv 4.2 \times 10^{-23}\,\text{kg}$.

$$T = 273 + 37 = 310\,\text{K}$$

$$\langle v^2\rangle^{1/2} = (3 \times 1.38 \times 10^{-23} \times 310/4.2 \times 10^{-23})^{1/2} = 17.4\,\text{m s}^{-1}$$

In solution, of course, the molecule doesn't travel very far at this speed before colliding with other (solvent) molecules and changing both speed and direction.

Don't forget to convert °C into K!

---

**Worked Problem 9.7**

**Q**: For a rotating object, rotational kinetic energy is $\frac{1}{2}(I)\omega^2$, where $\omega$ is the angular velocity of rotation and $I$ is the moment of inertia. What is the r.m.s. speed of thermal rotation at 37 °C for a spherical protein molecule of RMM 25 000 and radius 1.0 nm?

**A**: Mean thermal rotational energy $\frac{1}{2}I\omega^2 = 3kT/2$; therefore, r.m.s. rotational velocity, $\langle\omega^2\rangle^{1/2} = (3kT/I)^{1/2}$.
The moment of inertia of a uniform density sphere of mass $m$ and radius $r$, $I = 2mr^2/5$.
For the single protein molecule:

$$I = 2 \times 4.2 \times 10^{-23} \times (10 \times 10^{-10})^2/5 = 1.7 \times 10^{-41}\,\text{kg m}^2.$$

$$\langle \omega^2 \rangle^{1/2} = (3 \times 1.38 \times 10^{-23} \times 310/1.7 \times 10^{-41})^{1/2} 2.7$$
$$\times 10^{10} \text{rad s}^{-1} \equiv \text{ca. } 5 \times 10^9 \text{ rotations s}^{-1}.$$

Again, in solution this motion will be very jerky and severely damped by collisions with surrounding solvent molecules (see rotational diffusion, Section 4.5). However, this sort of tumbling frequency may be typical for macromolecules in a vacuum such as when being analysed in a mass spectrometer.

This sort of thermal translation and rotational motion is what is responsible for the random, chaotic Brownian motion observable in microscopic particles.

Thermodynamic fluctuations and Brownian motion can frustrate our attempts at understanding how enzymes and biological 'molecular machines' might work. It is now accepted, for example, that transmission of information through biological macromolecules (allostery, cooperativity) involves changes in dynamic fluctuations as well as simple conformational change.[1,2]

More generally, perpetual collisions with molecules in the surrounding environment give rise to **thermodynamic fluctuations**. For example, the mean square energy fluctuations of any object as it exchanges thermal energy to-and-fro with its surrounding are given by:

$$\langle \delta E^2 \rangle = kT^2 c$$

where $c$ is the heat capacity of the object. These energy fluctuations can be comparable to the free energy of folding for globular proteins.[1]

Most of the time, we don't see all this thermal chaos going on at the molecular level. All these fluctuations tend to cancel out when dealing with the large populations of molecules in typical experimental samples.

However, this raises the question: how valid is it to study the average effects of large populations of molecules when, in reality, they do their real job just one at a time? This is a classic dilemma in the theory of statistical thermodynamics, and the **ergodic hypothesis** is used to get around it. The ergodic hypothesis (or principle) argues that the average behaviour of a single particle over a sufficient length of time will be the same as the average of a large population of particles at any instant of time. This is actually difficult to prove, but it is a very useful hypothesis nonetheless. So, to the extent that this is true, do we really need to study single molecules? Well, maybe not—but it's fun to try anyhow.

## 9.3    Atomic Force Microscopy

Close your eyes and run your fingers over a surface. Even with your eyes closed you can feel the size, shape and texture of objects on the

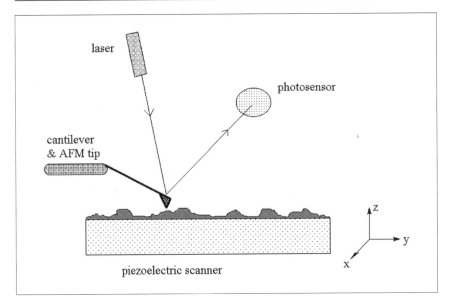

**Figure 9.1**  Atomic force microscopy.

surface. This is what atomic force microscopy (AFM) tries to do, only at a molecular level (Figure 9.1).

The 'finger' here is a very sharp tip that is mounted on the end of a flexible cantilever arm (usually made of silicon or silicon nitride). The position of the tip is measured by reflecting a beam of light onto a photodiode detector so that as the tip is dragged over the sample surface it moves up and down following the height (z) of the sample. By scanning in the x–y plane of the sample, a complete contour map of the sample surface can be generated. At the same time, bending of the flexible cantilever arm can be used to measure the strength of the forces involved as the tip interacts with the surface.

Relative movement of the AFM tip with respect to the sample along all three axes (x,y,z) is achieved by making use of the **piezoelectric effect.** A piezoelectric material is a substance (usually a crystal or ceramic material) that changes dimensions when a voltage is applied. Mounting the sample or cantilever arm onto a piezoelectric device allows either the tip or the sample support (or both) to be moved by tiny amounts, down to 0.1 nm or less in some instances, so that the very finest atomic/molecular surface detail can be probed. All this is done under very careful computer control so that the tip does not inadvertently push too hard and damage the very delicate surface it is trying to probe.

For simple imaging applications, the AFM tip may be dragged across the surface as described above (the so-called 'contact' or 'static' mode) or, alternatively, vibrated up and down near its resonant

The original AFM probes were made by gluing tiny slivers of diamond onto gold foil. Nowadays they are made on a larger scale by the same sort of microlithographic fabrication methods used for making computer silicon chips. Typical commercially available AFM tips are mounted on the end of a flexible silicon nitride cantilever, or 'arm', 100 μm long, 10 μm wide and ≈1 μm thick. The tip itself may be pyramidal or cone-shaped, with a tip radius of around 30 nm. Even finer (2–10 nm radius) and more durable AFM tips can be made using carbon nanotubes.

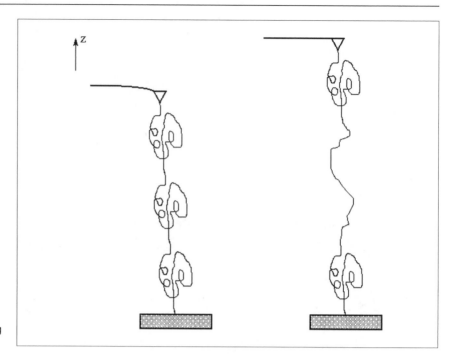

**Figure 9.2** Sketch illustrating steps in the mechanical unfolding of a tethered molecule.

frequency in the z-direction ('acoustic' or 'tapping' mode). The tapping mode gives additional information about the elastic and other mechanical properties of the material being scanned.

With suitably sharp AFM tips, almost atomic resolution can be achieved, at least in principle, and it has been used to determine the shapes of protein and nucleic acid molecules adsorbed on surfaces. One major advantage is that the method does work under water, so that molecules can be studied under near physiological conditions.

In more sophisticated experiments, the AFM tip can be used to grab hold of parts of a tethered molecule to see how it responds to being pulled apart. Alternatively, the forces between molecules adsorbed onto the AFM tip and those adsorbed onto the surface can be measured. This has been used, for example, to measure the forces involved in the mechanical unfolding of proteins, antibody–antigen and other protein–protein interactions, or the binding of complementary strands of DNA (Figure 9.2).

### Worked Problem 9.8

**Q**: In a typical AFM experiment to mechanically unfold a single globular protein, the cantilever arm moved by 25 nm with an average force of 150 pN as the polypeptide chain unravelled. How much mechanical work was done in the process? How does this compare to the Gibbs free energy of unfolding?

**A**: Work = Force × Distance = $150 \times 10^{-12}$ (N) × $25 \times 10^{-9}$ (m) = $3.75 \times 10^{-18}$ J (for one molecule).

This is equivalent to: $N_A \times 3.75 \times 10^{-18} = 2250 \, \text{kJ mol}^{-1}$

Under ideal, thermodynamically reversible conditions, this mechanical work is equal to the Gibbs free energy change ($\Delta G$) for the process. The numbers obtained here cannot be compared directly with the values obtained for molecules in solution using bulk thermodynamic methods (Chapter 5) because they include the additional elastic work involved in stretching the unfolded polypeptide as the tethered ends are pulled further apart than they would normally be for an unfolded protein free in solution.

Scanning tunnelling microscopy (STM) works in a rather similar way to AFM, but the surface is detected by measuring the very weak electric current that flows (by electron tunnelling) between the sample and the tip when they are very close together with a voltage between them. Unlike AFM, this cannot be done under water.

## 9.4   Optical Tweezers and Traps

One of the big problems in single molecule experiments is how to keep the molecule in place long enough to make observations on it, as well as how to grab hold of different parts of the molecule in order to manipulate it. In AFM and similar techniques, the molecules are immobilized by being adsorbed or attached on to macroscopic surfaces. Another method relies on the properties of finely-focussed laser beams to act as optical tweezers or optical traps to manipulate microscopic particles suspended in solution.

When a laser light beam is focused on a suspension of particles made of dielectric materials with a refractive index greater than the surrounding medium, the particles tend to congregate at the focus of the light beam. This is an effect arising from radiation pressure, as the change in momentum of the photons scattered from the particle creates tiny forces that can be sufficient to overcome the normal diffusion or Brownian motion of the particle. If the intensity of the light hitting the particle is different at different points on the object, then the object will move and—perhaps counter-intuitively—the motion tends to be towards the place where the light is most intense. The reason for this can be seen as follows.

Imagine a spherical particle suspended in a liquid (typically this might be a latex sphere or polystyrene bead a few microns in diameter in water). Because the refractive index of the particle is (usually) greater than the surrounding medium, beams of light passing through the object will be refracted, or change direction, as shown in Figure 9.3. Conservation of momentum requires that wherever the light beam (or stream of photons) changes direction, there will be an equal and opposite transfer of momentum to the particle. This occurs both on entry and exit of the beam, as shown in Figure 9.3, and results in forces (radiation pressure) on the particle with components both along the direction of the beam (x-axis) and perpendicular (y-axis).

Consider light beams A and B. If the light intensity is uniform across the beam, then the transverse radiation forces will tend to cancel, and there will be no overall transverse force on the particle (though it will tend to move along the x-axis). If, however, the light intensity is greater at A than at B, then the transverse forces will no longer cancel and the particle will move in the y-direction towards the region of greater intensity. This will normally be the centre of the laser beam. Moreover, if the laser beam is sharply focused, so that the light intensity varies along the x-axis as well, then there will be similar differential forces along the beam that will force the particle towards the focus of the beam where the light is most intense.

In this way the position of small particles can be manipulated with high precision over very small distances.

These sorts of optical tweezer techniques are also used to trap clouds of atoms in vapours, and to slow down their motion—effectively to cool them to very low temperatures—until they begin to exhibit Bose–Einstein condensation and other bizarre quantum effects.

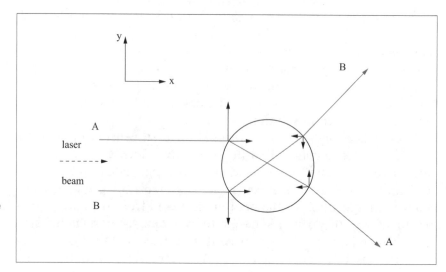

**Figure 9.3**  Ray paths for light beams passing through a transparent sphere, with vectors (arrows) indicating the transverse (y-axis) and longitudinal (x-axis) components of the forces generated at each optical interface.

This effect can be used to position and manipulate single cells of large viruses in solution. For other biophysical studies, individual macromolecules can be tethered to one or more latex or plastic beads and manipulated using optical tweezers. Other techniques combine similar approaches using small magnetic beads.

## 9.5   Single Molecule Fluorescence

Once you have pinned your molecule down, you need some way of observing what it gets up to. Fluorescence-based methods are usually the most sensitive here because photomultipliers and CCD devices (see Chapter 2) can readily detect the single photons emanating from individual fluorescent groups. Moreover, the intrinsic fluorescence of proteins or other fluorescence probes is very sensitive to conformational changes and is well understood for bulk samples.

With single molecules, however, the fluorescence emission is necessarily very weak—just a single photon (at most) from any excitation event.

---

### Worked Problem 9.9

**Q**: Imagine a fluorescent molecule with quantum efficiency ($\phi$) of 10% excited repetitively at one second intervals by an intense laser light pulse. What proportion of these pulses would result in a measurable emitted photon?

**A**: One in ten—at most. Even if the excitation pulse were intense enough to excite the molecule every time, there is only a 10% probability that the electronically excited state will decay radiatively. The rest of the time (90%) the decay will be non-radiative and the energy will dissipate as heat. (Note that fluorescence lifetimes are quite short and that decay back to ground state will be complete before the next excitation pulse arrives.)

---

The problem then arises: how do we know that a given photon comes from fluorescence of the single molecule and not from other sources of background light?

One approach (adopting methods already developed for more conventional samples) is to use time-resolved detection methods. Most of the stray photons from a laser pulse come from elastic (Rayleigh) or inelastic (Raman) scattering from the solvent and surroundings, and

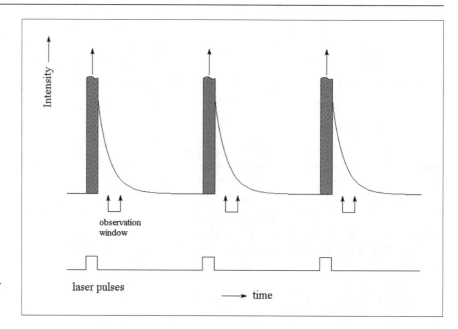

**Figure 9.4** Pulse sequence for time-resolved fluorescence measurement.

the intensity of this can far outweigh the light emitted from the (single) test molecule. However, such scattered light occurs essentially instantaneously compared with fluorescence emission which can take place over a longer (microsecond) timescale. Consequently, if we delay measuring photons until after the excitation pulse is completed, then we should only see the ones we want. This is the basis for numerous time-resolved methods, a simple version of which is shown in Figure 9.4.

Figure 9.4 illustrates the light intensity (fluorescence + scatter) as a function of time from a sequence of short excitation pulses. During the excitation pulse 'dead time' (shaded) the intense stray light background masks the fluorescence emission. However, if observation is delayed (electronically), then later stages of fluorescence emission may be observed free of this stray light contamination. This has the further advantage that, by varying the timing of the observation window, excited state lifetime properties can be measured. For single-molecule studies, data from a large number of pulses would be accumulated.

**Fluorescence correlation spectroscopy** is another technique that takes advantage of the sensitivity of laser-induced fluorescence methods to observe the diffusion and interaction of single molecules in a small volume of solution. The method is similar in principle to **dynamic light scattering** (see Section 4.6) and uses the fluctuations in fluorescence intensity from a sharply focused continuous laser beam. The volume of solution illuminated at the focus of a laser beam can be as little as

$10^{-15}\,dm^3$. As individual (fluorescent) molecules diffuse in or out of this volume element, the fluorescence intensity will fluctuate and analysis of the frequency spectrum (autocorrelations) within these fluctuations gives information about molecular size and diffusion dynamics.

### Worked Problem 9.10

**Q**: How many molecules are there in $10^{-15}\,dm^3$ of a 1 nmol $dm^{-3}$ solution?

**A**: $10^{-15}\,dm^3 \times 10^{-9}\,moles/dm^3 \times N_A = 0.6$, *i.e.* roughly one per volume element.

An advantage of this technique is that the single molecules do not have to be tethered (unlike, for example, AFM), but are free in solution. Interactions between single molecules or conformational changes (*e.g.* unfolding) might be detected by changes in the diffusion dynamics.

### Summary of Key Points

1. Single molecules are very dynamic and difficult to pin down.
2. Atomic force microscopy and optical tweezer techniques can be used to manipulate single macromolecules.
3. Atomic force microscopy can also be used to obtain low resolution images of single macromolecules.
4. Fluorescence is one of the more sensitive optical properties suitable for single molecule detection.

### Problems

**9.1.** What is the effective molar concentration for one molecule contained within:

(a) a box of side $10\,\mu m$;

(b) an icosahedral virus capsid (internal diameter $\approx 30\,nm$);

(c) a $C_{60}$ fullerene (cavity diameter $\approx 0.7\,nm$);

(d) the focal volume of a sharply focussed laser beam ($\approx 10^{-15}\,dm^3$)?

**9.2.** Estimate the root mean square (r.m.s.) energy fluctuations at room temperature in a protein molecule of RMM 25 000. (The specific heat of protein is typical of organic matter at about $3\,J\,K^{-1}\,g^{-1}$.)

**9.3.** What would such energy fluctuations correspond to in terms of temperature fluctuation?

**9.4.** If AFM is so good and can achieve atomic resolution, why is it not used to determine the three-dimensional structures of proteins or nucleic acids?

**9.5.** Why is it necessary to use repetitive excitation pulse sequences to measure single-molecule fluorescence properties?

## References

1. A. Cooper, Protein fluctuations and the thermodynamic uncertainty principle, *Prog. Biophys. Mol. Biol.*, 1984, **44**, 181–214.
2. A. Cooper and D. T. F. Dryden, Allostery without conformation change, *Eur. Biophys. J.*, 1984, **11**, 103–109.

## Further Reading

A. Ashkin, Optical trapping and manipulation of neutral particles using lasers, *Proc. Natl. Acad. Sci. U. S. A.*, 1997, **94**, 4853–4860.

C. Bustamante, S. B. Smith, J. Liphardt and D. Smith, Single-molecule studies of DNA mechanics, *Curr. Opin. Struct. Biol.*, 2000, **10**, 279–285.

C. Bustamante, Z. Bryant and S. B. Smith, Ten years of tension: single-molecule DNA mechanics, *Nature*, 2003, **421**, 423–427.

J. N. Forkey, M. E. Quinlan and Y. E. Goldman, Protein structural dynamics by single-molecule fluorescence polarization, *Prog. Biophys. Mol. Biol.*, 2000, **74**, 1–35.

M. S. Z. Kellermayer, S. B. Smith, H. L. Granzier and C. Bustamante, Folding-unfolding transitions in single titin molecules characterized with laser tweezers, *Science*, 1997, **276**, 1112–1116.

J. M. Sneddon and J. D. Gale, *Thermodynamics and Statistical Mechanics,* RSC Tutorial Chemistry Text, Royal Society of Chemistry, Cambridge, 2001, ch. 9.

P. J. Walla, *Modern Biophysical Chemistry: Detection and Analysis of Biomolecules*, Wiley-VCH, Weinheim, 2009.

J. Zlatanova, S. M. Lindsay and S. H. Leuba, Single molecule force spectroscopy in biology using the atomic force microscope, *Prog. Biophys. Mol. Biol.,* 2000, **74**, 37–61.

# Answers to Problems

---

## Chapter 1

**1.1.** Number of protein molecules in $1\,cm^3 = (45 \times 10^{-3}/65\,000) \times N_A = 4.2 \times 10^{17}$

Volume per molecule $= 1/4.2 \times 10^{17} = 2.4 \times 10^{-18}\,cm^3$

Average distance apart $= (2.4 \times 10^{-18})^{1/3} = 1.3 \times 10^{-6}\,cm\,(13\,nm)$

Mass of one molecule $= 65\,000/6 \times 10^{23} = 1.1 \times 10^{-19}\,g$

This corresponds to a molecular volume of around $1.1 \times 10^{-19}\,cm^3$, assuming a density similar to water. This corresponds to a cube of side approximately 4.8 nm (or about 6 nm diameter if we assume the molecule is spherical, with volume $4\pi r^3/3$).
So, in a $45\,mg\,cm^{-3}$ solution, these molecules are separated on average by about 2–3 molecular diameters.

**1.2.** (a) Assume that, because of stereochemical constraints, each $\phi$ and each $\psi$ angle can adopt three possible orientations ($120°$ apart), so that statistically each peptide unit might have $3 \times 3 = 9$ possible conformers.
For 100 peptide units, the number of possible conformers $= 9^{100} = 2.7 \times 10^{95}$.
[Note: You may find that your calculator gives an error message when you try to calculate numbers such as $9^{100}$ directly. If so, try breaking it down into several steps, for example: $9^{50} \times 9^{50}$.]
(b) $1\,fs = 10^{-15}\,s$

$$\text{Time taken} = (\text{number of conformers}) \times (\text{time per}$$
$$\text{conformer})$$
$$= 2.7 \times 10^{95} \times 10^{-15}\,s = 2.7 \times 10^{80}\,s$$
$$\approx 9 \times 10^{72}\,\text{years}$$
$$(1\,\text{year} \approx 30\,\text{million seconds})$$

For comparison, the estimated age of the Universe is about 15 gigayears ($1.5 \times 10^{10}$ years).

(This is one version of the 'Levinthal Paradox' and the protein folding computational problem.)

**1.3.** $\Delta E = mgh$ (gravitational potential energy)

$$h = \Delta E / mg = \Delta E / (70 \times 9.81) = \Delta E / 686.7$$

(a) $\Delta E = 10\,g \times 17\,kJ\,g^{-1}$ (for carbohydrate) $= 170\,000\,J$ $(\Delta E) \rightarrow$ $h = 250\,m$ (sugar)
(b) $\Delta E = 10\,g \times 39\,kJ\,g^{-1}$ (for fats) $= 390\,000\,J$ $(\Delta E) \rightarrow h = 570\,m$ (fat)

The calculation assumes that all the metabolic energy is used just for the work done in raising the gravitational potential energy of the 70 kg weight. Actual climbing (or jumping) is less mechanically efficient.

**1.4.** $7000/39 = 180\,g$ per day

**1.5.** Liquid water is most dense at $4\,°C$ and will sink to the bottom; any ice or water not at $4\,°C$ will be less dense and will float above this.

**1.6.** (a) Relatively high melting and boiling points, density increases on melting (ice floats), maximum density at $4\,°C$, high heat capacity for the liquid, high dielectric constant, *etc.*
(b) Hydrogen bonding in water gives rise to the open, low density, tetrahedral structure of ice without which icebergs would sink rather than float (and therefore not be a potential hazard to ships). Furthermore, hydrogen-bonded solids are quite strong, giving ice the strength to crush or penetrate ships hulls.

**1.7.** (a) $q_{Na+} = -q_{Cl-} = 1.6 \times 10^{-19}$ C; $\varepsilon_r = 1$ for vacuum;

$$r = 5\,Å = 0.5\,nm = 5 \times 10^{-10}\,m$$

$$V_{qq} = q_1 q_2 / 4\pi\varepsilon_0\varepsilon_r r = -(1.6 \times 10^{-19})^2 /$$
$$(4\pi \times 8.85 \times 10^{-12} \times 5 \times 10^{-10})$$
$$= -4.6 \times 10^{-19}\,J \text{ per ion pair (note minus sign,}$$
$$\text{attractive interaction)} \times N_A$$
$$\equiv -276\,kJ\,mol^{-1}$$

(b) In water, $\varepsilon_r \approx 80$: $V_{qq} = -3.5\,kJ\,mol^{-1}$

This is comparable to thermal energy [$kT$ (per molecule) $\equiv RT$ (per mole) $\approx 2.5\,kJ\,mol^{-1}$]. This is why salts dissolve readily in water.

**1.8.**  (a) Increased thermal motion will tend to disrupt the re-orientation of dipolar molecules in an electric field, with consequent reduction in $\varepsilon_r$ at higher temperature.

(b) The electrostatic attraction between oppositely charged groups in water will get stronger as the temperature increases ($\varepsilon_r$ is in the denominator of the Coulomb potential, $V_{qq}$ becomes more negative as $\varepsilon_r$ decreases).

(c) Endothermic ($\Delta H$ positive). The Coulomb potential ($V_{qq}$) is the work done, or free energy change, in bringing the two charges together from an infinite separation. Applying the Le Chatelier's principle, any interaction that gets more favourable with increase in temperature should be endothermic. This is counterintuitive, but can be rationalized at the molecular level in terms of the (endothermic) release of solvation molecules around the individual charges as they are brought together. The release of these solvent water molecules is entropically favourable ($\Delta S$ positive), so the overall free energy change is attractive ($\Delta G = \Delta H - T\Delta S$ is negative).

**1.9.**   Various possible answers ...

(i) Popular newspapers can't get their sums right.
(ii) The calculation in Worked Problem 1.4 assumed an average base pair separation appropriate for double-helical DNA. This distance will be longer for fully unfolded single-strand DNA.
(iii) Three billion base pairs would be six billion nucleotides if the complementary strands were separated into single chains.

## Chapter 2

**2.1.**   (a) Thermal atomic/molecular vibrations, black body radiation

(b) Electronic transitions in gas, fluorescence in tube coating material

(c) Oscillations of free electrons in a cavity (magnetron)

(d) Inner-shell electron transitions induced by collision of high energy electrons with a metal target

(e) Stimulated electronic transitions in atoms or molecules

(f) Acceleration of high energy electron beam (or other charged particle) around a circular path

(g) Radio frequency oscillations in electronic circuit.

**2.2.**   (a) 0.1  (10%);  (b)  0.01  (1%);  (c)  $10^{-5}$  (0.001%); (d) 1 (100%) $[T = 10^{-A}]$

**2.3.**   (a)  2, 1.3, 0.6, 0.3, 0.046, 0; (b) 0.63, 0.32, 0.1, 0.03, $1 \times 10^{-10}$

**2.4.** 0.996, 1.96, 2.7

Calculation: actual transmission of the sample is $T\% = 100 \times 10^{-A} = 10.0$, 1.0 or 0.1%, respectively, for each of the samples. However, '0.1% stray light' means that an additional 0.1% of light reaches the detector, so the apparent $T\% = 10.1$, 1., or 0.2%. Apparent (measured) $A = -\log_{10}(T\%/100) = 0.996$, 1.96 or 2.7.

**2.5.** Very little light passes through high absorbance samples, so stray light makes a proportionally great contribution to the light intensity at the detector.

**2.6.** $\varepsilon_{280} = 5690 \times n_{Trp} + 1280 \times n_{Tyr} + 60 \times n_{Cys}$ (half cysteines only).
Phe does not absorb significantly at 280 nm. No information is given for half cysteines (usually insignificant).

Lysozyme : $\varepsilon_{280} = (6 \times 5690) + (3 \times 1280) = 37980$
(predominantly Trp)
Insulin : $\varepsilon_{280} = (0 \times 5690) + (4 \times 1280) = 5120$ (only Tyr)
Ribonuclease : $\varepsilon_{280} = (0 \times 5690) + (6 \times 1280) = 7680$ (only Tyr)
Albumin : $\varepsilon_{280} = (1 \times 5690) + (18 \times 1280) = 28730$
(mainly Tyr)

**2.7.** Various possibilities, including:

- protein not all correctly folded;
- impure preparation—contains some other UV absorbing impurity (*e.g.* other proteins or DNA contaminant from preparation);
- incorrect sequence used for cloning;
- missing co-factor.

**2.8.** Dye binding, amino acid analysis after total hydrolysis, total nitrogen determination (though none of these will tell whether the protein is correctly folded).

**2.9.** By comparison with Figure 2.21, ABA-1 is predominantly α-helical and RS is mainly β-sheet.

**2.10.** Amino acids (except for glycine) contain a chiral centre ($C_{\alpha}$) and are intrinsically optically active; predominantly L-amino acids in protein, with no racemization during (gentle) hydrolysis. Purine and pyrimidine bases are achiral, because they have no asymmetric carbons or other chiral centres.

**2.11**. Synthetic mixtures are racemic.

**2.12**. Frank–Condon principle (see Figure 2.22). The electronic excited state lifetime is usually long enough for the system to relax to the excited state energy minimum before de-excitation occurs. Consequently, no matter how the excited state is created, emission will always involve the same (vertical) transition, with the same energy.

**2.13**. Absorbance $A = -\log(T)$; $T = 10^{-A} = 10^{-0.1} = 0.79$ for a 0.5 cm path length.

(Remember we are focusing on the centre of the fluorescence cuvette.) Consequently, with this absorbance, about 80% of the exciting light reaches (or leaves) the middle of the sample solution. This is (just about) acceptable. Any higher absorbance would lead to significant loss of light and distortion of the observed spectrum.

**2.14**. $kT = 1.38 \times 10^{-23} \times 300 = 4.2 \times 10^{-21}$ J at room temperature.

Vibrational band energy, $\delta E = hc/\lambda$

Wavenumber, $1/\lambda = \delta E/hc = 4.2 \times 10^{-21}/(6.626 \times 10^{-34} \times 3 \times 10^{8})$
$= 21\,000\,\text{m}^{-1} = 210\,\text{cm}^{-1}$

**2.15**. $285\,\text{nm} \equiv 35\,100\,\text{cm}^{-1}$, $318\,\text{nm} \equiv 31\,400\,\text{cm}^{-1}$

Therefore $\Delta v = 3700\,\text{cm}^{-1}$ (–OH overtone region).

## Chapter 3

**3.1**. Adjacent peaks (usually) correspond to different protonation states ($\pm 1\,\text{H}^{+}$) of the acidic and basic residues.

**3.2**. From eqn (3.5):

$$v = (2zeV/m)^{1/2} = (2eV/1\,\text{amu})^{1/2}(z/m)^{1/2}$$
$$= 1.96 \times 10^{6}(z/m)^{1/2}\,\text{m s}^{-1}$$

(a) $z/m = 1$; $v = 1.96 \times 10^{6}\,\text{m s}^{-1}$
(b) $z/m = 1/132$; $v = 1.7 \times 10^{5}\,\text{m s}^{-1}$
(c) $z/m = 4/14500$; $v = 3.3 \times 10^{4}\,\text{m s}^{-1}$

**3.3**. $\text{TOF} = \text{distance/velocity} = 1.5/v$

(a) $0.76\,\mu\text{s}$; (b) $8.8\,\mu\text{s}$; (c) $45\,\mu\text{s}$

**3.4**. Use eqn (3.4): $r = (2mV/zeB^{2})^{1/2}$

(a) For $m=1$ amu, $z=1$, $r=(2 \times 1.66 \times 10^{-27} \times 20\,000/1.6 \times 10^{-19} \times 4^2)^{1/2} = 5.1 \times 10^{-3}$ m (0.51 cm). Other ions are proportional to $(m/z)^{1/2}$: (b) 5.8 cm; (c) 30.7 cm.

**3.5.** One complete (circular) orbit comprises $2\pi$ radians, or a distance (circumference) of $2\pi r$. Consequently, the angular frequency (radians per second) for an object moving with velocity, $v$, is $\omega = 2\pi \times$ number of orbits per second $= 2\pi \times (v/2\pi r) = v/r$.
Eqn (3.2) $(r = mv/zeB)$ rearranges to give $v/r = zeB/m$.

**3.6.** At least two possible approaches using mass spectrometry:

(a) Differences in RMM for the protein dimer $(2 \times 13\,700 = 27\,400)$ compared to the fusion protein $(13\,700 + 12\,500 = 26\,200)$ should be detectable by MS (*e.g.* MALDI-TOF).
(b) Alternatively, peptide mass fingerprinting after proteolysis of the impurity should show the presence (or absence) of known GST peptides.

## Chapter 4

**4.1.** For a two-component system, total volume $V = \bar{v}_1 g_1 + \bar{v}_2 g_2$
and density, $\rho = (g_1 + g_2)/V = (g_1 + g_2)/(\bar{v}_1 g_1 + \bar{v}_2 g_2)$
so that (rearrange): $\bar{v}_2 g_2 = (g_1 + g_2)/\rho - \bar{v}_1 g_1$
For pure water (component 1), the density of $0.99707$ g cm$^{-3}$ means that:
$\bar{v}_1 g_1 = \bar{v}_1 \times 0.99707 = 1.0000$ cm$^{-3}$, or the partial specific volume of water, $\bar{v}_1 = 1/\rho_1 = 1.002939$ cm$^3$ g$^{-1}$
For the protein solution, $\rho = 0.99748$ g cm$^{-3}$
$\bar{v}_2 g_2 = (g_1 + g_2)/\rho - \bar{v}_1 g_1 = (5.0 + 0.0075)/0.99748 - (1.002939 \times 5.0)$
$= 0.005456$ cm$^3$
$\bar{v}_2 = 0.005456/0.0075 = 0.727$ cm$^3$ g$^{-1}$

**4.2.** Repeat calculation (answer 4.1) with $\rho = 0.99752$ g cm$^{-3}$
$\bar{v}_2 g_2 = (5.0 + 0.0075)/0.99752 - (1.002939 \times 5.0) = 0.005254$ cm$^3$
$\bar{v}_2 = 0.005254/0.0075 = 0.701$ cm$^3$ g$^{-1}$
This corresponds to a volume decrease of around 4% upon unfolding.

**4.3.** There are several possible reasons:

(i) The folded protein structure may contain voids or cavities due to imperfect packing that might become occupied by solvent water when the structure unfolds.

(ii) Changes in extent and structure of hydration/solvation layers upon unfolding.

[Note: in practice, protein volumes are seen to both increase and decrease upon unfolding, depending on the specific protein and experimental conditions.]

**4.4.** An unbalanced rotor causes excessive vibrations and mechanical stresses while spinning that can cause breakage.

**4.5.** The rotational kinetic energy in a spinning rotor (see below) can cause injury if the rotor becomes detached.

**4.6.** $m = 2\,kg$; $r = 0.15\,m$; $\omega = 2\pi \times 40\,000/60 = 4200\,rad\,s^{-1}$. Rotational kinetic energy $\approx \frac{1}{2}\,mr^2\omega^2 = \frac{1}{2} \times 2 \times (0.15)^2 \times (4200)^2 = 4 \times 10^5\,J$

This corresponds to the explosive energy in about 87 g of TNT.

**4.7.** Use $D = RT/6\pi N_A \eta R_S$ (Stokes $-$ Einstein equation)

$$= 8.314 \times 293/(6\pi \times 6 \times 10^{23} \times 1.002 \times 10^{-3} \times R_S)$$

$$= 2.15 \times 10^{-19}/R_S\,m^2\,sec^{-1}$$

(a) $R_S = 0.5\,nm$, $D = 2.15 \times 10^{-19}/0.5 \times 10^{-9} = 4.3 \times 10^{-10}\,m^2\,sec^{-1}$

(b) $R_S = 2.5\,nm$, $D = 2.15 \times 10^{-19}/2.5 \times 10^{-9} = 8.6 \times 10^{-11}\,m^2\,sec^{-1}$

(c) $R_S = 5\,\mu m$, $D = 2.15 \times 10^{-19}/5 \times 10^{-6} = 4.3 \times 10^{-14}\,m^2\,sec^{-1}$

**4.8.** Use $\langle x^2 \rangle = 6Dt$ (where $\langle x^2 \rangle$ is the mean square displacement in any direction), with $t = 300\,s$.

$$\text{Root mean square displacement, } x_{rms} = \langle x^2 \rangle^{1/2} = (6Dt)^{1/2}$$
$$= (1800D)^{1/2}$$

(a) $x_{rms} = (1800 \times 4.3 \times 10^{-10})^{1/2} = 8.8 \times 10^{-4}\,m = 0.88\,mm$
(b) $x_{rms} = (1800 \times 8.6 \times 10^{-11})^{1/2} = 1.5 \times 10^{-7}\,m = 150\,nm$
(c) $x_{rms} = (1800 \times 4.3 \times 10^{-14})^{1/2} = 8.8 \times 10^{-6}\,m = 8.8\,\mu m$

**4.9.** Thermal velocities, calculated from the thermal kinetic energy, $\frac{1}{2}mv^2 \approx 3kT/2$ (Chapter 5), assume no collisions with surrounding molecules.

## Chapter 5

**5.1.** Use $\frac{1}{2}m\langle v^2 \rangle = 3kT/2$, $\langle v^2 \rangle = 3kT/m = 3 \times 1.381 \times 10^{-23} \times T/m$ where $m$ is the mass of a single molecule (in kg).

(a) RMM of $O_2 = 32$; molar mass $= 32 \times 10^{-3}\,\text{kg mol}^{-1}$; $T = 298\,\text{K}$; $m = 32 \times 10^{-3}/N_A = 5.3 \times 10^{-26}\,\text{kg}$

$$\langle v^2 \rangle = 3kT/m = 3 \times 1.381 \times 10^{-23} \times 298/5.3 \times 10^{-26}$$
$$= 2.3 \times 105\,\text{m}^2\,\text{s}^{-2}$$

Root mean square velocity, $\langle v^2 \rangle^{1/2} = 482\,\text{m s}^{-1}$

(b) RMM of $H_2O = 18 \times 10^{-3}\,\text{kg mol}^{-1}$; $T = 298\,\text{K}$; $\langle v^2 \rangle^{1/2} = 642\,\text{m s}^{-1}$

(c) Protein: $m = 25\,000 \times 10^{-3}\,(\text{kg mol}^{-1})/N_A = 4.2 \times 10^{-23}\,\text{kg}$; $T = 310\,\text{K}$; $\langle v^2 \rangle^{1/2} = 17.6\,\text{m s}^{-1}$

**5.2.** Atmospheric pressure is what we feel as a result of high-speed molecular collisions with the air molecules in the atmosphere around us. Brownian motion also shows the effects of molecular collisions on the (random) motion of small particles.

**5.3.** (a) Use $\Delta G° = -RT.\ln K = \Delta H° - T.\Delta S°$ to complete the table:

| t/°C | K | $\Delta G°$/kJ mol$^{-1}$ | $\Delta H°$/kJ mol$^{-1}$ | $\Delta S°$/J K$^{-1}$ mol$^{-1}$ |
|---|---|---|---|---|
| 45 | 0.133 | 5.33 | 150.0 | 454.9 |
| 50 | 0.345 | 2.86 | 175.0 | 532.9 |
| 55 | 1 | 0 | 200.0 | 609.8 |
| 60 | 3.22 | −3.24 | 225.0 | 685.4 |

(b) Fraction unfolded $= K/(1 + K) = 0.26, 0.5, 0.76$ (respectively)

(c) Increase in $\Delta H°$ with temperature (confirmed by increase in $\Delta S°$ with $T$) signifies a positive heat capacity increment ($\Delta C_p$), characteristic of both hydrophobic and hydrogen-bonded network interactions.

**5.4.** Spectroscopic methods (UV, fluorescence, CD)—changes in chromophore environment/conformation on binding.
Hydrodynamics (viscosity, sedimentation)—changes in gross macromolecular properties.
Calorimetry (DSC, ITC)—direct measure of energy changes on binding.
Equilibrium dialysis—direct measure of ligand binding.

**5.5.** Both F and CD follow the same transition in this case, so may use either.

(a) $T_m = 50\,°C$ (mid-point of unfolding transition)

(b) Fraction unfolded $= (F - F_0)/(F_{inf} - F_0) = (58.8 - 65)/(15 - 65) = 0.124$

(c) Equilibrium constant for unfolding:

$$K = (F - F_0)/(F_{inf} - F) = (58.8 - 65)/(15 - 58.8) = 0.142$$
$$\Delta G_{unf}^{\circ} = -RT.\ln K = -8.314 \times (273 + 46) \times \ln(0.142)$$
$$= +5.18 \, \text{kJ} \, \text{mol}^{-1}$$

(d) Fluorescence is probing the polarity of the environment of aromatic amino acid residues (primarily tryptophan), which changes (non-polar → polar) as the protein unfolds. CD measures secondary structure changes (α-helix, β-sheet, *etc.*).

(e) These do not necessarily occur simultaneously, since protein unfolding may take place in two (or more) steps, *e.g.* change in tertiary structure exposing aromatic groups but retaining secondary structure, followed by 'melting' of the secondary structure at higher temperatures.

**5.6.**                              $$P + L \rightleftharpoons PL$$

$$\therefore c_p/[PL] = 1 + [P]/[PL] = 1 + 1/K[L]$$

Slope of 'double-reciprocal plot' (1/[PL] *versus* 1/[L]) $= 1/Kc_p$. Useful because only equilibrium dialysis (and related methods) gives free ligand concentration [L] directly. For most other methods we need to make approximations, or fit to complete binding expression.

**5.7.** $c_p = [PL] + [P] = 8.3 \times 10^{-9}$ M (from left hand compartment)

$c_L = [PL] + [L] = 3.9 \times 10^{-8}$ M  ...

Free ligand $[L] = 3.5 \times 10^{-8}$ M (from right hand compartment)

Therefore $[PL] = 3.9 \times 10^{-8} - 3.5 \times 10^{-8} = 4.0 \times 10^{-9}$ M

$$[P] = 8.3 \times 10^{-9} - 4.0 \times 10^{-9} = 4.3 \times 10^{-9} \, \text{M}$$

Hence $K = [PL]/[P][L] = 2.7 \times 10^7 \, \text{M}^{-1}$

## Chapter 6

**6.1.** Use $A_{diffusion} = 4\pi N_A \, r_{XY}(D_X + D_Y) \times 1000$

Assume target does not diffuse ($D_Y = 0$), $D_X = 10^{-10} \, \text{m}^2 \, \text{sec}^{-1}$, $r_{XY} = 1$ nm

$A_{diffusion} = 4\pi \times 6 \times 10^{23} \times 10^{-9} \times 10^{-10} \times 1000 = 7.5 \times 10^8 \, \text{M}^{-1} \, \text{s}^{-1}$

**6.2.** $\text{Rate} = A_{\text{diffusion}}[X] = 7.5 \times 10^8 \times 1 \times 10^{-6} = 750 \, \text{s}^{-1}$

$t_{1/2} = 0.693/k \approx 0.693/750 = 9 \times 10^{-4} \, \text{s} \, (0.9 \, \text{ms})$

**6.3.** Reaction is not under free diffusion control; or there is pre-formed reactive complex; or the diffusion is two-dimensional (on surface) or one-dimensional (along a polymer strand); or molecular attractions are enhancing collision rates; or the assumed mechanism is wrong; or there is an unexpected catalytic effect; . . . and so on.

**6.4.** $6 \times 10^{-12} \, \text{s} \times 3 \times 10^8 \, \text{m s}^{-1} = 0.0018 \, \text{m} \, (0.18 \, \text{cm})$

**6.5.** (a) Observed rate is proportional to peptide concentration.
(b) $k_{\text{on}} = \text{rate}/[\text{peptide}] = 1.9 \times 10^7 \, \text{mol}^{-1} \, \text{dm}^3 \, \text{s}^{-1}$ (same for all concentrations)
(c) $t_{1/2} = 0.693/k = 7.2 \, \text{s}; \, k_{\text{off}} = 0.693/t_{\frac{1}{2}} = 0.096 \, \text{s}^{-1}$
(d) $K = k_{\text{on}}/k_{\text{off}} = 1.9 \times 10^7/0.096 = 2.0 \times 10^8 \, \text{mol}^{-1} \, \text{dm}^3$

## Chapter 7

**7.1.** Affinity chromatography, using a nickel column (binds histidine side chains); elute with a pH gradient or imidazole.

**7.2.** Gel filtration/size exclusion chromatography using a column calibrated with known proteins in the 10 000–50 000 RMM range. If the protein is a dimer (30 000), it will elute sooner than expected from the column.

**7.3.** Electrophoresis is only partly helpful here. If the dimer is held together by covalent bonds (*e.g.* –S–S– bridges under non-reducing conditions), then SDS-PAGE would show a protein band at around 30 000. However, non-covalent dimerization of the native state would be disrupted by SDS and only monomer bands would be seen.

**7.4.** Reverse phase/hydrophobic interaction chromatography. The contaminating fatty acids should have a greater affinity for the hydrophobic groups on the column and the protein will pass through, leaving the fatty acids behind.

## Chapter 8

**8.1.** The function of haemoglobin/myoglobin is to transport oxygen, so it is important that oxygen molecules can have access to the heme binding sites, which are buried deep inside the globular protein structures. So it was initially surprising to find the path blocked—no obvious routes or channels for oxygen to diffuse in or out of the heme binding pocket. Some possibilities:

(i)   The crystal structure is wrong? Always possible, but the structure has subsequently been solved by other groups using other methods (*e.g.* neutron diffraction) and the result is always the same. This lack of access is found also in other proteins.

(ii)  Crystal packing distorts the protein conformation? Is the structure in solution the same as in a crystal? This has been much discussed, but there is little evidence to suggest that crystal packing forces have any serious effect on protein structure. Protein crystals contain a lot of water (40% or more?), with relatively few protein–protein contacts. Comparisons with NMR structures in solution rarely show any significant differences.

(iii) Protein dynamics? This is the most likely explanation. X-ray crystallography gives a (mostly) static, averaged view of the protein conformation. In reality, each molecule is undergoing thermal motion/conformational fluctuations, during which temporary channels can open and close to allow passage of small molecules (see Sections 2.2.4, 2.4.3 and 9.2).

Reference: M. F. Perutz, Myoglobin and haemoglobin: the role of distal residues in reaction with haem ligands, *Trends Biochem. Sci.*, 1989, **14**, 42–44.

**8.2.** Use Bragg's law (Box 8.1), *i.e.* $2d.\sin\theta = n\lambda$

Scattering angle $2\theta = 26.2$; $\theta = 13.1°$

Spacing $d = n\lambda/2.\sin\theta$

$$= 1.5418/(2.\sin13.1), \text{i.e. assuming first order}, n = 1$$

$$= 3.4 \text{ Å} (0.34 \text{ nm}).$$

(Note: if you get a different answer, check that your calculator is set to degrees (not radians, *etc.*) for trigonometric functions.)

This repeat corresponds to the base pair separation in double-helical DNA.

**8.3.** Kinetic energy: $\frac{1}{2}mv^2 = 3kT/2$ (see Section 5.1)

$$v = (3kT/m)^{1/2} = (3 \times 1.381 \times 10^{-23} \times 293/1.67 \times 10^{-27})$$
$$= 2696 \text{ m s}^{-1}$$

de Broglie wavelength: $\lambda = h/mv = 6.626 \times 10^{-34}/(1.67$
$$\times 10^{-27} \times 2696)$$
$$= 1.47 \times 10^{-10}\,\text{m}(0.147\,\text{nm}, 1.47\,\text{Å})$$

(h is Planck's constant, $6.626 \times 10^{-34}\,\text{Js}$)

## Chapter 9

**9.1.** Molar concentration for one molecule in volume $V$ (in $\text{dm}^3$) $= 1/N_A V$ ($1\,\text{m}^3 = 1000\,\text{dm}^3$)

(a) $V = (10 \times 10^{-6})^3 \times 1000 = 1 \times 10^{-12}\,\text{dm}^3$;
$c = 1.7 \times 10^{-12}\,\text{mol}\,\text{dm}^{-3}$
(b) $V = 4\pi r^3/3 \approx 1.4 \times 10^{-23}\,\text{m}^3 \equiv 1.4 \times 10^{-20}\,\text{dm}^3$;
$c = 1.2 \times 10^{-4}\,\text{mol}\,\text{dm}^{-3}$
(c) $V = 4\pi r^3/3 \approx 1.8 \times 10^{-28}\,\text{m}^3 \equiv 1.8 \times 10^{-25}\,\text{dm}^3$;
$c = 9.3\,\text{mol}\,\text{dm}^{-3}$
(d) $V = 10^{-15}\,\text{dm}^3$; $c = 1.7 \times 10^{-11}\,\text{mol}\,\text{dm}^{-3}$

**9.2.** Use $\langle \delta E^2 \rangle = kT^2 c$ where $c$ is the heat capacity of the single molecule.

For a single molecule, $c \approx 25\,000 \times 3/N_A = 1.25 \times 10^{-19}\,\text{J}\,\text{K}^{-1}$

$\langle \delta E^2 \rangle$; $= 1.38 \times 10^{-23} \times 300^2 \times 1.25 \times 10^{-19} = 1.55 \times 10^{-37}\,\text{J}^2$

$\delta E_{rms} = 3.9 \times 10^{-19}\,\text{J per molecule} \equiv 240\,\text{kJ}\,\text{mol}^{-1}$.

**9.3.** $\delta T = \delta E/c \approx 3.9 \times 10^{-19}/1.25 \times 10^{-19} \approx 3\,°\text{C}$

**9.4:** AFM probes only the surface and cannot give details of internal conformations. It probes only the top surface and cannot feel underneath.

**9.5.** Because: the single molecule may only fluoresce a small fraction ($\phi$) of the time; or the (single photon) emission may not always occur within the observation time; or the emitted photon may come off in the wrong direction for the detector.

# Subject Index